Powder and Bulk Solids
Handling Processes

CHEMICAL INDUSTRIES

A Series of Reference Books and Textbooks

Consulting Editor
HEINZ HEINEMANN
Heinz Heinemann, Inc.,
Berkeley, California

Additional Volumes in Preparation

Powder and Bulk Solids Handling Processes

INSTRUMENTATION AND CONTROL

Koichi Iinoya

*The Society of Powder
 Technology, Japan
Kyoto, Japan*

Hiroaki Masuda

*Hiroshima University
Higashi-Hiroshima, Japan*

Kinnosuke Watanabe

*Sankyo Dengyo Co., Ltd.
Tokyo, Japan*

CRC Press
Taylor & Francis Group
Boca Raton London New York

CRC Press is an imprint of the
Taylor & Francis Group, an informa business

First published 1988 by Marcel Dekker, Inc.

Published 2019 by CRC Press
Taylor & Francis Group
6000 Broken Sound Parkway NW, Suite 300
Boca Raton, FL 33487-2742

© 1988 by Taylor & Francis Group, LLC
CRC Press is an imprint of Taylor & Francis Group, an Informa business

First issued in paperback 2019

No claim to original U.S. Government works

ISBN 13: 978-0-367-45128-8 (pbk)
ISBN 13: 978-0-8247-7971-9 (hbk)

Visit the Taylor & Francis Web site at
http://www.taylorandfrancis.com

and the CRC Press Web site at
http://www.crcpress.com

Library of Congress Cataloging-in-Publication Data

Iinoya, Koichi.
    Powder and bulk solids handling processes : instrumentation
and control / Koichi Iinoya, Hiroaki Masuda, Kinnosuke
Watanabe.
        p.   cm. -- (Chemical industries; v. 34)
        Includes bibliographies and index.
        ISBN 0-8247-7971-1
        1. Bulk solids handling. I. Masuda, Hiroaki. II.
Watanabe, Kinnosuke. III. Title. IV. Series.
    TS180.8.B8136 1988
    620'.43--dc19   88-14195
                                                              CIP

Preface

Powders and bulk solids are found in various industries as raw
materials or as intermediate or final products. The synthetic-
energy production industry often starts with a solid feed such
as coal, oil shale, or industrial waste. The pharmaceutical industry manufactures powders and bulk solids as the final products. The ceramics industry treats powders very carefully,
because the powders should be kept free of contamination.
Powder and bulk solids handling is therefore an important operation in many industries, including ferrite manufacturing, cement,
fertilizer, pigment, nuclear fuel, and steel. The more sophisti-
cated the processes are, the more important control of powder
and bulk solids handling processes becomes.

    Powder handling is difficult compared to fluid handling.
Whereas funding of the study of fluids has been very extensive,
less has been available for the study of powders and bulk solids.
Consequently, systematic studies of the powder process instrumentation have been sparse in the past. Also, powder handling
processes have been left uncontrolled, whereas chemical industries

based on petroleum have been fully developed. In these situations, instrumentation of the powder and bulk solids handling
processes will have a direct impact on various industries in the
very near future.

   The main purpose of this book is to reveal some key points
in the instrumentation of powder handling processes and to provide readers with sufficient knowledge of instrumentation to act
as a base for the development of new instruments. For practic-
ing engineers and scientists, the book should be helpful for
understanding and applying measurement principles in their daily
work. It should also be suitable as a textbook for an advanced
undergraduate or graduate course. The basics of powder technology are also given so that readers unfamiliar with these subjects can study them easily.

   In the first chapter we list related industrial fields and the
physical representation of powder properties, as an introduction
to the book. In Chapter 2 we introduce instrumentation, including a brief review of state variables in powder processes and
physical or chemical phenomena applicable to the instrumentation.
Readers are provided with an outline of modern instrumentation.
Chapter 3 concerns the sampling of particles from a powder bed
or suspension flow as a basic method for evaluating powder handling processes. Collecting a representative sample of particles
may be difficult without a thorough knowledge of particle behavior. The eight chapters that follow represent various methods
and instruments for on-line measurement of powder flow rate,
particle concentration in suspension, level of powder in storage
vessels, particle size, moisture content of powder, powder pressure and fluid pressure, temperature, and slurry viscosity. Instruments for the measurement of the quality of powder mixtures
are also discussed in the last chapter. Readers familiar with
powder processes can start with Chapter 4. The physical and
chemical principles in each chapter are described as simply as
possible. References are included with each chapter so that the
ideas introduced can be explored in more detail.

Koichi Iinoya
Hiroaki Masuda
Kinnosuke Watanabe

Contents

    3.1   Powder Sampling   39
    3.2   Sampling of Gas-Solid Suspensions   51
          References   62

4   Powder Flow Rate   65

    4.1   Introduction   65
    4.2   Weighing Method   67
    4.3   Flowmeters Based on Mechanics   73
    4.4   Differential Pressure Method   79
    4.5   Electrical Method   87
    4.6   Statistical Method   99
    4.7   Optical Method and Others   102
    4.8   Flow Detectors   105
          References   107

5   Particle Concentration in Suspensions   109

    5.1   Introduction   109
    5.2   Definition of Particle Concentration   111
    5.3   Electrical Method   113
    5.4   Attenuation Technique   119
    5.5   Aerosol Concentration   124
          References   129

6   Level of Powder in Storage Vessels   131

    6.1   Introduction   131
    6.2   Mechanical Method   132
    6.3   Electrical Method   138
    6.4   Ultrasonic Wave Level Meter   142
    6.5   Radiometric Method   144
    6.6   Pneumatic Method and Others   147
          References   149

7   Particle Size   151

    7.1   Introduction   151
    7.2   Direct Classification Method   153
    7.3   Permeability Method   159
    7.4   Optical Method   164
    7.5   Electrical Method   173
    7.6   Acoustic Method and Others   178
          References   193

Contents

Powder and Bulk Solids Handling Processes

1

Introduction to
Powder Handling Processes

There are many types of industries where powder is produced
by chemical reactions or handled as raw or intermediate materials.
Here we give a brief review of various industries relating to powder handling processes. The physical representation of powder
properties is then explained as a basis of the process instrumentation. Finally, some examples of powder handling processes
are discussed.

1.1 RELATED INDUSTRIAL FIELDS

Powder handling processes play an essential role in the following
industries, among others [1]:

Cement	Pigment
Pharmaceuticals	Fertilizer
Steel	Cosmetics
Paper	Nuclear fuel

Detergent	Powder metallurgical
Refractory material manu- facturing	Feed production
	Synthetic energy
Whetstone and abrasives	Ferrite manufacturing
Explosives production	Toner production
Flour milling	New ceramics

Operations in powder handling processes are classified into two categories. One is mechanical, such as size reduction of raw materials, filtration and separation of fine particles, size classification, powder mixing, and granulation. The second, based on heat and mass transfer, includes drying, crystallization, and extraction. Each is called a unit operation. Chemical industries are composed of unit processes for which unit operations are carried out. The several types of mechanical unit operations and their roles in industrial processes are summarized briefly below.

Size reduction: crushing and grinding of solids or particles so as to obtain particles that meet size specifications. The purposes are to make it easy to handle the solid materials in the processes that follow, to make it easy to separate various components contained in the solid, to obtain a higher reaction rate or activity of the solid, to control product quality, and so on.

Gas-solid separation: to recover the particles from gas-solid suspension as a process product or to clarify the process gas. Dust collection in soot and smoke-emitting facilities is an important operation to prevent air pollution.

Filtration: to collect particles from liquid-solid suspension in order to recover the particles as a process product or to clarify the liquid.

Classification: to classify particles according to their properties, such as particle diameter, particle shape, density, electric conductivity, magnetism, and so on. Size classification is one of the most widely used unit operations.

Powder mixing: to distribute more than two types of powder so as to get homogeneous powder of multiple components.

Mulling: to coat the surface of particles uniformly with liquid or paste. Mulling is an operation essential to the plastic treatment of powder.

Granulation: to produce particles of controlled size and shape from powder, liquid suspension, or fused material. Powder handling in subsequent processes becomes easier through granulation. Granulation is also required for the products to fit various applications.

Storing: to store powder in a storage vessel. The purpose is to keep a large amount of powder as feed, intermediate, or final material in industrial processes. Storage vessels work both as the buffer volumes in the processes and as protectors against contamination or change in powder quality.

Transportation: to transport powder from storage vessels, storage yards, or powder handling equipment to subsequent powder handling processes.

Feeding: to feed a specified amount of powder, at a specified rate, into subsequent powder handling equipment, such as chemical reactors, mixers, classifiers, granulators, and so on.

Expression: to squeeze out a liquid component from biological particles or powder accompanied by liquid, such as filtered cakes. Expression is a well-known unit operation in wine-making facilities.

Plastic treatment: to make various objects of different shapes from powder. Before the plastic treatment, small amounts of plasticizer and binder will be added to the powder.

Calcination: thermal treatment of powder or various objects obtained through the plastic treatment. Particles will be sintered through the calcination and the objects become hard and strong. Electrical properties of ceramics, for example, are controlled by calcination.

The above-mentioned unit operations are best combined in the chemical industries, where plastics, paint, pigment, fertilizer, and other items are produced. Based on their products, the chemical industries are also classified as plastics industries, paint manufacturing industries, fertilizer industries, and so on.

In the powder metallurgical industries, there are also important powder handling operations, such as size reduction, size classification, powder mixing, plastic treatment of powder, and calcination. Filtration, dewatering, and drying are also necessary in wet processes producing metallic powder, such as electrolyte copper powder. Through these processes, the industries make many important products, such as ceramic tools, hard metals, gears, cermet tools, or oil-less bearings. Development of the ceramic engine is one of the most recent goals of the powder metallurgical industries. The quality of metallurgical products depends greatly on the amount of impurities and additives, the mean particle size, and the size distribution of pulverized raw materials.

There are many useful electronic ceramics, such as piezo-electric elements, ceramic capacitors, electrical resistance elements, optoelectronic elements, and ferrite devices. These ceramics are produced through processes similar to those in metallurgical industries. The production of uranium dioxide (UO_2) pellets for use as nuclear fuel is composed of similar unit processes.

Other industries, such as pharmaceutical, feed production, food production, and cosmetics, utilize powder as their raw or intermediate material and make various types of products: medical tablets, agricultural chemicals, instant coffee, and lipstick. The blending or granulation process in these industries must be carefully controlled because the products might be harmful if the component ratios differed from the specifications. A variation in the water content of raw powders brings about product deterioration. Further, the contamination of different types of powders and contamination by bacteria or foreign substances should be completely avoided in these industries. The health effects of pharmaceutical powder on industrial workers should also be considered, and well-designed dust-collecting systems should be introduced.

Powder handling is a necessary operation in the paper industry, where fine powders such as talc are utilized to produce smooth-surfaced papers on which high-density printing is possible. The rubber industry utilizes very fine carbon black powders. For example, tires for racing cars are produced from a number of powders blended at specified ratios that affect the lap time in auto races.

As briefly reviewed, the powder handling processes play an important role in many industries. It may be that higher-order problems will be brought about as products of higher quality are required. Further research will be required to solve such problems. Considerable research is now going on in the following fields:

Agricultural engineering   Chemical engineering
Civil engineering   Electrical engineering
Food engineering   Mechanical engineering
Pharmaceutical technology   Powder technology

1.2 PARTICLE SIZE

The size of a spherical particle is uniquely represented by its
diameter. Particles are not always spherical, however, and there
are several ways of defining particle diameter, as noted in Table
1.1. Geometrical representation of a nonspherical particle is

Table 1.1   Terms Relating to Particle Size

Term	Note
Breadth, length, thickness	b, ℓ, t (dimensions of circumscribed rectangular parallelepiped)
Arithmetic mean diameter of the three dimensions	$(b + \ell + t)/3$
Geometric mean diameter of the three dimensions	$(b\,\ell\,t)^{1/3}$
Harmonic mean of the three dimensions	$3/(1/b + 1/\ell + 1/t)$
Projected area equivalent diameter (Heywood diameter)	$\pi D_p^2/4 = A$ (A; projected area)
Volume equivalent diameter (equivalent volume diameter, sphere equivalent diameter)	$\pi D_e^3/6 = V_p$ (V_p, particle volume)
Feret diameter (Green diameter)	Unidirectional particle diameter
Martin diameter	Unidirectional diameter equally bisecting the projected area
Others	Stokes diameter, sieve diameter, optical diameter, electrical mobility diameter

based on the three dimensions of a circumscribed rectangular parallelepiped. One definition is given by the following equation:

$$D_p = \frac{\ell + b + t}{3}$$ (1.2.1)

where ℓ is the length, b the breadth, and t the thickness of the circumscribed rectangular parallelepiped. The particle diameter represented by Eq. (1.2.1) is called the arithmetic mean diameter. It is important to distinguish the mean diameter from the mean particle diameter described in Section 1.4. The arithmetic mean diameter is a definition of a particle size based on a single particle, whereas the mean particle diameter is a physical representation of the average size of particles constituting the powder. The geometric mean diameter based on the three dimensions is also one of the definitions of particle size.

   There are several definitions based on the projected shape of a particle. They are: the diameter of a circle of equal projected area (projected area diameter or Heywood diameter), the diameter of a circle of equal perimeter (perimeter diameter), the Feret (or Green) diameter, and the Martin diameter. The Feret diameter and the Martin diameter are defined as follows;

Feret diameter: unidirectional diameter obtained as the distance
    between pairs of parallel tangents to the projected figure of
    the particle
Martin diameter: unidirectional diameter equally bisecting the
    projected area

   Both the Feret diameter and Martin diameter are statistical, because each of the particles will take random orientations in the measurements. The following experimental relationship was found by Heywood [2]:

   Feret diameter > projected area diameter > Martin diameter

                                                            (1.2.2)

There is a mathematical relation between the projected area A averaged for all orientations and the surface area S of a particle if the particle is convex (Cauchy's theorem):

$$S = 4A \qquad\qquad (1.2.3)$$

For a sphere, A equals $(1/4)\pi D_p^2$ and S is given by πD_p^2.

Another type of definition of particle diameter is based on the concept of an equivalent sphere. The volume equivalent diameter D_e is defined by the following relationship:

$$\frac{\pi}{6}D_e^3 = V_p \qquad\qquad (1.2.4)$$

where V_p is the particle volume. For example, the volume equivalent diameter of an aggregate of n equal spheres of size D is given by $\sqrt[3]{n}\, D$.

The surface equivalent diameter and the specific surface equivalent diameter are defined similarly. These equivalent diameters are called the sphere equivalent diameters. The diameter of a sphere having settling velocity equal to that of the actual particle is also one of the equivalent diameters. This is called the Stokes diameter for a particle sufficiently small that the fluid drag on the particle is given by Stokes' law.

1.3 PARTICLE SIZE DISTRIBUTION

Powder contains very many particles of different sizes. The size distribution of particles represents one of the most important physical properties of powder.

1.3.1 Cumulative Distribution and Frequency Distribution

Sieve analysis and many other methods give the cumulative size distribution as an oversize or undersize distribution. The following relationship exists between the undersize distribution $F(D_p)$ and the oversize distribution $R(D_p)$:

$$F(D_p) = 1 - R(D_p) \qquad\qquad (1.3.1)$$

The frequency distribution of particle size $f(D_p)$ is obtained by differentiating the cumulative distribution:

$$f(D_p) = \frac{dF}{dD_p} = -\frac{dR}{dD_p} \tag{1.3.2}$$

Whereas the cumulative distribution is dimensionless, the frequency distribution has the unit meter^{-1}. The following equation will also be obtained by Eq. (1.3.2):

$$F(D_p) = \int_0^{D_p} f \, dD_p \tag{1.3.3}$$

The frequency distribution of particles should satisfy the following equation:

$$\int_0^\infty f \, dD_p = 1 \tag{1.3.4}$$

That is, the frequency distribution should be normalized.

Powder that has a wide size distribution is said to be polydisperse; powder that has a very narrow size distribution is called monodisperse powder. Ideally, monodisperse powder is composed of equal-sized particles, and the frequency distribution is represented by a unit impulse function.

1.3.2   Count Base Distribution and Mass Base Distribution

The count base distribution is obtained by counting the number of particles, as in the case of microscopic analysis or the Coulter counter measurement. The mass base distribution is obtained by weighing the particles as in sieve analysis or sedimentation balance measurement. The count base distribution does not coincide with the mass base distribution.

If particles are of the same shape and density, the count base distribution $f^{(0)}$ can be transformed into the mass base distribution $f^{(3)}$ by use of the following equation:

$$f^{(3)} = \frac{D_p^3 f^{(0)}}{\int_0^\infty D_p^3 f^{(0)} \, dD_p} \tag{1.3.5}$$

Equation (1.3.5) can be generalized as follows:

$$f^{(\beta)} = \frac{D_p^{\beta-\alpha} f^{(\alpha)}}{\int_0^\infty D_p^{\beta-\alpha} f^{(\alpha)} \, dD_p} \qquad (1.3.6)$$

where α and β take the following integer, depending on the measurement base:

−1: specific surface base
0: count base
1: length base
2: area base
3: mass base

1.3.3   Representation of Particle Size Distributions

It is well known that many types of statistical data are represented by the Gaussian distribution (also called the normal distribution), where the frequency distribution is symmetrical about the mean. The particle size distribution is, however, not well represented by the distribution. The two distributions described below are most widely utilized to represent the particle size distributions.

Log-Normal Distribution

The distribution is called log-normal if the logarithm of particle size is represented by the normal distribution. That is,

$$R(\ln D_p) = \frac{1}{\sqrt{2\pi}\,\sigma} \int_{\ln D_p}^\infty \exp\left[-\frac{(\ln D_p - \mu)^2}{2\sigma^2}\right] d(\ln D_p)$$

$$(1.3.7)$$

where the two parameters μ and σ are the mean and the standard deviation of the log-normal distribution, respectively.

$$\mu = \ln D_{p50} \qquad (1.3.8)$$

$$\sigma = \ln \sigma_g \qquad\qquad\qquad (1.3.9)$$

D_{p50} is the particle size where R takes the value 0.5. This is called the median size. For the log-normal distribution, D_{p50} is equal to the geometric mean diameter D_g of the powder. On the other hand, σ_g is the geometric standard deviation given by the following equation:

$$\sigma_g = \frac{D_{p15.9}}{D_{p50}} = \sqrt{\frac{D_{p15.9}}{D_{p84.1}}} \qquad\qquad (1.3.10)$$

Equation (1.3.10) can be derived by use of the characteristics of the Gaussian distribution [e.g., $R(\mu + \sigma)$ of the Gaussian distribution is 0.159].

The corresponding frequency distribution $f(\ln D_p)$ is usually given by $-dR/d(\ln D_p)$. The frequency distribution thus obtained is dimensionless, because $d(\ln D_p) = dD_p/D_p$ is dimensionless.

Rosin-Rammler or Weibull Distribution

The following equation will fit better than the log-normal distribution for powders of broader size distribution.

$$R(D_p) = \exp\left[-\left(\frac{D_p}{D_{p36.8}}\right)^n\right] \qquad\qquad (1.3.11)$$

This is also a two-parameter distribution, as in the case of the log-normal distribution. The parameter $D_{p36.8}$ is the diameter of R = 0.368 (=1/e), as is easily recognized from Eq. (1.3.11). On the other hand, the parameter n represents the sharpness of the distribution. The larger the parameter n, the narrower the distribution is.

The frequency distribution is given by the following equation:

$$f(D_p) = \frac{n D_p^{n-1}}{D_{p36.8}^n} \exp\left[-\left(\frac{D_p}{D_{p36.8}}\right)^n\right] \qquad\qquad (1.3.12)$$

From Eq. (1.3.12), it is easily shown that $f(0) \neq 0$ for $n \leq 1$.

1.4 MEAN PARTICLE DIAMETER

The median diameter is one type of mean particle diameter. The median diameter on the mass base $D_{p50}^{(3)}$ is called the mass median diameter (MMD), and $D_{p50}^{(0)}$ on the count base is called the count median diameter (CMD) or the number median diameter (NMD). The geometric mean diameter D_g is also one type of mean particle diameter. These mean particle diameters are often utilized in an analysis of powder handling processes.

However, various phenomena observed in powder handling processes will depend not only on the mass median diameter but also on the particle size distribution. Different powder will give a different result even though the median diameter is kept constant throughout an observation. In this section we discuss the mean particle diameter, which should be utilized in an analysis of powder handling processes.

In a determination of mean particle diameter, it is assumed that there are no mutual interactions between particles. Let $y(D_p)$ represent the particulate process. Then the observed value of the particulate process $\overline{y(D_p)}$ is given by the following equation:

$$\overline{y(D_p)} = \int_0^\infty y(D_p)f(D_p)\ dD_p \qquad (1.4.1)$$

The mean particle diameter $\overline{D_p}$ is determined so that the observed value is given by $y(\overline{D_p})$. That is,

$$y(\overline{D_p}) = \overline{y(D_p)} \qquad (1.4.2a)$$

or

$$\overline{D_p} = y^{-1}(\overline{y}) \qquad (1.4.2b)$$

where y^{-1} is the inverse function of y.

Example 1.4.1 Find the mean particle diameter of the process proportional to D_p^2, such as the Stokes sedimentation of particles.

Solution   From Eq. (1.4.1),

$$\overline{y(D_p)} = k \int_0^\infty D_p^2 f(D_p) \, dD_p \tag{a}$$

From the definition (1.4.2),

$$k \, \overline{D_p}^2 = k \int_0^\infty D_p^2 f(D_p) \, dD_p \tag{b}$$

Therefore,

$$\overline{D_p} = \sqrt{\int_0^\infty D_p^2 f(D_p) \, dD_p} \tag{c}$$

When the mean particle diameters determined by Eq. (1.4.2) have the same value for different powders, the linear parts of the particulate process have the same value even if the size distribution parameters of these powders are not the same. Therefore, there may be no scattering in the data due to inadequate selection of the mean particle diameter. Thus the various data measured may be systematically analyzed only by use of the definition (1.4.2). Further discussion on this subject may be found in the literature [3].

The following mean particle diameters are often utilized in an analysis of powder handling processes:

Mean surface diameter:

$$D_s = \sqrt{\int_0^\infty D_p^2 f^{(0)} \, dD_p} \tag{1.4.3}$$

Mean volume diameter:

$$D_v = \sqrt[3]{\int_0^\infty D_p^3 f^{(0)} \, dD_p} \tag{1.4.4}$$

Sauter diameter (mean volume surface diameter):

$$D_{sv} = \frac{\int_0^\infty D_p^3 f^{(0)} \, dDp}{\int_0^\infty D_p^2 f^{(0)} \, dDp} \tag{1.4.5}$$

The Sauter diameter is the same as the harmonic mean diameter on the mass base:

$$D_{sv} = D_h^{(3)} = \frac{1}{\int_0^\infty (f^{(3)}/D_p) \, dD_p} \tag{1.4.6}$$

If the particles are spherical, the harmonic mean diameter can be obtained by the following equation:

$$D_h^{(3)} = \frac{6}{S_v} \quad \text{for spherical particles} \tag{1.4.7}$$

where S_v is the specific surface area (m^2/m^3). The numerical constant 6 in Eq. (1.4.7) is the specific surface shape factor for spherical particles. If the specific surface area S_v is obtained through an adequate measurement, $D_h^{(3)}$ will be determined by Eq. (1.4.7) even if the particles are not spherical. The mean particle diameter thus determined is called the specific surface area diameter.

Numerical integration may be necessary to obtain the numerical value of the mean particle diameter derived by the definition (1.4.2). The calculation is greatly simplified by the following three lemmas when the size distribution of the powder is lognormal:

Lemma 1:

$$\int_{-\infty}^\infty D_p^m f^{(\beta)} (\ln D_p) \, d(\ln D_p) = \exp\left(\mu^{(\beta)} m + \frac{1}{2} m^2 \sigma^2 \right)$$

$$\tag{1.4.8a}$$

Lemma 2:

$$\bar{y}^{(\beta)}(\mu^{(0)}, \sigma^2) = \bar{y}^{(\alpha)}(\mu^{(0)} + (\beta - \alpha)\sigma^2, \sigma^2) \qquad (1.4.8b)$$

Lemma 3:

$$\mu^{(\beta)} = \mu^{(\alpha)} + (\beta - \alpha)\sigma^2 \qquad (1.4.8c)$$

The following type of mean particle diameter is easily calculated by the use of Lemma 1:

$$\bar{D}_p^{(\beta)} = \left[\int_{-\infty}^{\infty} D_p^m f^{(\beta)}(\ln D_p)\, d(\ln D_p) \right]^{1/m}$$

$$= D_{p50}^{(\beta)} \exp\left(\frac{1}{2} m \ln^2 \sigma_g\right) \qquad (1.4.9)$$

The count base data are transformed into the mass base by the use of Lemma 2.

$$\bar{y}^{(3)}(\mu^{(0)}, \sigma^2) = \bar{y}^{(0)}(\mu^{(0)} + 3\sigma^2, \sigma^2) \qquad (1.4.10)$$

The count median diameter is transformed into the mass median diameter by the use of Lemma 3.

$$D_{p50}^{(3)} = D_{p50}^{(0)} \exp(3 \ln^2 \sigma_g) \qquad (1.4.11)$$

Example 1.4.2   How can the mean particle diameter in Example 1.4.1 be represented on a mass base for the powders of log-normal size distribution?

Solution   From Eq. (1.4.9) as m = 2,

$$\bar{D}_p^{(3)} = D_{p50}^{(3)} \exp(\ln^2 \sigma_g) \qquad (a)$$

1.5 EXAMPLES OF POWDER HANDLING PROCESSES

1.5.1 Coal Combustion

Figure 1.1 shows a process in which pulverized coal is produced as the fuel for an industrial process such as cement production. The raw coal stored in a silo is fed continuously to a mill, where the coal is pulverized to achieve a suitable size distribution. The pulverized coal is transported by the flow of hot air from the mill outlet to a size classifier, where larger particles are separated and returned to the mill inlet. Then the air suspension of finer particles is further transported to a cyclone separator placed above a feed hopper, and there the coal particles are recovered into the hopper. The coal is stored in the feed hopper for 4 to 8 hours of continuous firing. The hopper is provided with feeders for volumetric or gravimetric coal-fuel feeding to the burners of a rotary kiln. The coal mill is dimensioned to produce the necessary quantity of pulverized coal for continuous firing of the kiln.

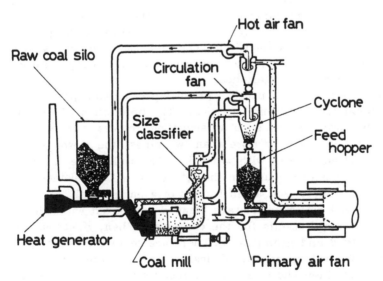

Figure 1.1 Coal combustion process. (Courtesy of F. L. Smidth, Copenhagen, Denmark.)

A circulation fan removes the air from the top of the cyclone. On the pressure side of the fan, the air is distributed through three branches, one leading to the suction side of the primary air fan, the other back to the size classifier, and the third branch returning the air to the mill inlet, where it is again used for mill ventilation. With this recirculating system, it is possible to attain the necessary air velocities through the mill and the cyclone separator irrespective of the primary air consumption for the kiln.

The method of coal combustion described above is called indirect firing, because the pulverized coal is stored in the feed hopper before firing. The advantages of indirect firing are effective control of coal-fuel feeding and a certain independence between the kiln and the mill. The firing can be continued as long as the coal fuel remains in the feed hopper, even if the mill is brought to a standstill.

1.5.2 New Ceramics Production

The raw materials for new ceramics, for example, Al_2O_3 (alumina), ZrO_2 (zirconium oxide), Si_3N_4 (silicon nitride), and SiC (silicon carbide), are produced only through chemical synthesis, because they do not exist as natural resources. These raw materials are used in the form of fine powder. The properties of the powder, such as the chemical composition, crystal form, density, particle shape, and particle size, are very important in getting ceramics with the desired characteristics.

Figure 1.2 shows a process producing calcinated powder as a raw material for electronic ceramics. The production of titanate or ferrite powder needs such a calcination process. In products consisting of powders of more than two types, the powders are blended and mixed well with water in a ball mill. The slurry is purified and then calcinated in the rotary kiln. The calcinated materials are pulverized again by use of a hammer mill.

Pulverized raw materials are blended and mixed with additives to meet the requirements for the ceramics. Then, if necessary, they are dried and granulated. There are several types of plastic treatment for making objects of desired shape. When these objects are sintered, they show various characteristics of new ceramics or electronic ceramics.

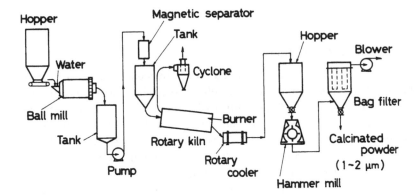

Figure 1.2   Calcinated powder production process (raw materials production for new ceramics).

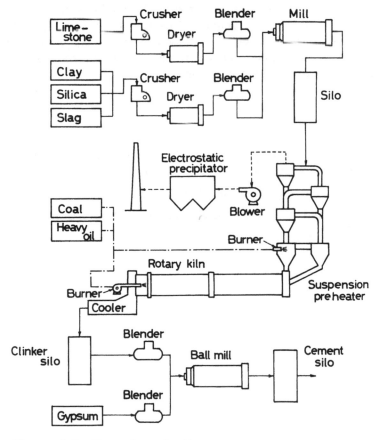

Figure 1.3   Cement production process.

1.5.3 Cement Production

Figure 1.3 shows the cement production process. The raw materials, such as limestone, silica, clay, and slag, are roughly pulverized and blended. The intermediate materials are pulverized again. The size reduction of particles in this stage is carried out carefully because the degree of calcination depends strongly on particle size. The feed materials for the second mill are thus well dried, so that the mill works smoothly. The mill feed rates should also be controlled; otherwise, it is difficult to meet the size specifications. The output powder of the mill is stored in a silo or a feed hopper.

The pulverized raw material is fed to the rotary kiln through the suspension preheaters and calcinated, producing cement clinkers. Cement is obtained by pulverizing the clinkers after blending with a small amount of gypsum. The size reduction process is usually formed by a closed-circuit system consisting of feeders, mills, classifiers, and gas-solid separators, as in the case of the coal combustion process shown in Fig. 1.1.

REFERENCES

1. K. Iinoya (ed.), Funtai Kogaku Binran (Handbook of Powder Technology), Nikkan Kogyo, Tokyo, part VII (1986).

2. H. Heywood, Chem. Ind. 56, 149 (1937).

3. H. Masuda and K. Iinoya, Mem. Fac. Eng. Kyoto Univ. 34, 334 (1972).

2

Introduction to Instrumentation

Instrumentation is one of the most important subjects in powder handling processes. For accurate control of a process, it is essential to know the operating conditions. There may be many variables that represent process situations, including the flow rate of powders, and the temperature and pressure of powder handling equipment. Several instruments are required for the measurement of these process variables. In this chapter we present a brief review of the variables that must be measured in the powder handling processes. The instruments themselves are described in Chapters 4 to 11.

One of the problems discussed here relates to the necessary conditions that instruments should satisfy to meet the process requirements. In Section 2.3 we review the various physical or chemical phenomena applicable to the instrumentation.

2.1 STATE VARIABLES IN POWDER
    HANDLING PROCESSES

As we have seen in Section 1.1, a powder handling process is composed of several subprocesses where the following unit operations are carried out:

1. Storing the particulate raw or product materials in storage vessels
2. Feeding the material into reactors or other particulate equipment
3. Transportation of the particulate material between pieces of equipment
4. Primary operations, such as grinding, classification, and separation (filtration, dust collection)

High quality products or highly controlled products can be obtained only through a high degree of control at every stage of the entire process. The first step is to measure the variables that represent the dynamic behavior of the process. The variables, which are called state variables for a process system, are functions of time. The important state variables in a powder handling process are listed below:

Powder Flow Rate, W (kg/s)

Powder flow rate is the mass of particles flowing through a certain area per unit time. The flow rate into equipment such as a reactor or separator controls the performance of the equipment. This state variable relates to most operations, including transportation, feeding, grinding, classification, and granulation. In Chapter 4 we describe the various methods for detecting the powder flow rate.

Particle Concentration in Suspensions, c (kg/m^3)

This variable controls, for example, the combustion of finely pulverized coal. The performance of gas-solid separation also depends strongly on the concentration. The powder flow rate through a unit cross section is given as a product of the concentration and the velocity. The variable relates to pneumatic

conveying, slurry transportation, dust collection, particle classification, gas-solid reaction, and so on. In Chapter 5 we describe various methods for measuring particle concentration.

Level of Powder in a Storage Vessel, L (m)

This variable indicates the physical situation for raw or product
materials stored in vessels. It also relates to feeding or discharging of materials. A blending process of several powders
fed from storage vessels depends strongly on the variable. If
one of the vessels happens to be empty, the blended products
have no value. Various types of level meters are discussed in
Chapter 6.

Particle Size, D_p (m)

This is the most important state variable representing a quality
of a particulate product. In coal combustion, particle size controls the efficiency of combustion. The variable also relates to
the cement production process, production of raw materials for
ceramics, crystallization, particle reaction, drying, plastic treatment of powder, classification, grinding, granulation, dust collection, filtration, and so on. The size required to be measured
is as small as 100 nm (= 10^{-7} m) in large-scale-integration (LSI)
manufacturing factories. In Chapter 7 we discuss related methods.

Moisture Content, w (%)

Moisture content is an especially important factor in the quality
control of grains, cereals, and other biological particles. The
variable relates to drying, storing, food production, and coal
combustion. In Chapter 8 we describe the measurement methods.

Pressure or Load, P [N/m^2 (= Pa)]

There are two types of pressure: powder pressure and fluid
(gas or liquid) pressure. Powder pressure relates to storing,
feeding, and transportation of solid materials. Storage vessels
are sometimes destroyed because of excessive powder pressure.
Therefore, pressure is an important state variable in the safety
supervision of storage vessels.

Fluid pressure relates to the transportation of solids by gas
or liquid flow. Efficiencies of such unit operations as dust collection, particle classification, and gas-solid reaction are also
associated with gas pressure. The energy or power consumption
of gas flow depends on the product of the pressure loss and the
flow rate. In Chapter 9 we discuss these state variables.

Temperature, T (K)

Many processes, including drying, combustion, calcination, and
chemical reaction, depend on the temperature of the process materials or surrounding gas. In Chapter 10 we describe various
thermometers.

Viscosity of Slurry, μ [Ns/m^2 (= Pa$\cdot$s)]

Viscosity is the most important property in the transportation of
slurry, such as a coal-water mixture (CWM). The variable also
relates to the dewatering and expression of slurry and to the
pharmaceutical production process. In Chapter 11 we describe
related methods of measurements.

Composition or Component, X (%)

Composition of powder is essential in pharmaceutical, agricultural-
chemical production, food production, and various blending processes. With a little change in composition, medicine may act as
a poison. Therefore, measurement of the composition should be
accurate. In these situations, laboratory measurements based on
off-line operations such as x-ray fluorescence are usual after the
process particles have been sampled. The sampling methods are
discussed in Chapter 3. On-line measurement of the degree of
mixing of different types of particles is discussed in Chapter 11.

Gas Flow Rate, Q (kg/s), Q_v (m^3/s)

Gas flow rate is one of the variables in control of the performance
of a process with gas-solid two-phase flow, such as pneumatic
conveying, dust collection, particle classification, and particle
grinding by air jet. The energy consumption of such a process

depends on the gas flow rate. The effect of suspended particles on the measurement of gas flow rate is discussed in Section 4.4.

We discuss next the coal combustion process shown in Fig. 1.1. The first problem is the quality and quantity control of the coal mill product. The operating condition of the mill must be kept under supervisory control. The important state variables in this case are the particle size and the flow rate of pulverized coal. The level of coal in the silo and the level of the pulverized coal in the feed hopper are also state variables essential in maintaining normal operation of a coal mill. Fine particles such as pulverized coal may, however, cause the following problems:

1. Particle bridging in the hopper
2. Particle adhesion to the wall and formation of a dead zone in the particle flow
3. Flushing of powder through a feeder or the hopper outlet

These phenomena must be considered when designing the control system.

Other problems in the coal combustion process are kiln control and waste gas control. Further instrumentation is required for the measurement of gas temperature (at mill and kiln outlets), coal feed rate to the kiln, and the moisture content of pulverized coal. The coal feed rate controls the temperature distribution and the main reaction in the kiln. Burners for the pulverized coal need powder flow meters for gas-solid suspensions. The moisture content affects the total coal consumption, because the vaporization of water contained in the coal wastes a large amount of the combustion heat.

Automatic operation of a powder handling process requires various powder feeders along with instruments for measuring the state variables listed earlier. Sometimes, the feeder discharges no particles even if the motor is rotating, because of particle adhesion, caking, pinching, and clogging. The movement of the main part of the feeder and the temperature of the motor are therefore the additional state variables that have to be monitored to avoid these problems.

2.2 NECESSARY CONDITIONS FOR INSTRUMENTS

To control a powder handling process, state variables must be transformed into the corresponding electric or pneumatic signals, based on the suitable physical phenomena. The transformed signals will be amplified and transmitted to a control station and the values indicated on a control panel or recorded on paper. Transformation of the state variables into electric signal (pneumatic or oil pressure signals) is called the first transformation, and the devices for the first transformation are called sensors. Amplification of the signal is the second transformation, and the final transformation is an output transformation for indication or recording. An instrument with these three types of transformation function is desirable. However, our purpose is to introduce the method of first transformation for powder handling processes.

A device for the first transformation, called a sensor, should satisfy the following conditions:

1. It has good repeatability and is easily calibrated.
2. It covers a wide range of measurement, or it is easy to change its range.
3. It has good sensitivity and is not affected by variables other than the objective state variable.
4. The signal is large enough and is not affected by disturbances or noise.
5. The dynamic response is as fast as possible.
6. The signal is linear; that is, the signal is proportional to the state variable. The signal has to be a monotonic function of the state variable, even if the linear condition is not fulfilled.
7. The characteristics do not change with the passage of time, and are chemically and physically stable. The sensors have to be rigid and have good resistance for abrasive wear, even if they contact particles.
8. Handling and maintenance are easy.

In relation to the first condition, we will discuss briefly the measurement error of a sensor. Here we denote the measured value by x_i, which is a discrete value obtained by the ith meas-

urement.   If we repeat the measurement on a unique true value
x' of a state variable, the data are usually represented by a
normal distribution with mean $\bar{x}$ and the standard deviation s
(see also Section 3.1.2).   The difference $(x' - \bar{x})$ is called the
bias of the sensor.   We can eliminate the bias through calibration.

The standard deviation s, which shows the degree of random
error, is defined by the following equation:

$$s = \sqrt{\frac{1}{n} \sum_{i=1}^{n} (x_i - \bar{x})^2} \qquad (2.2.1)$$

where n is the total number of data.   Then 68% of the data will
be found in the range $(\bar{x} - s)$ to $(\bar{x} + s)$.   The range and the
fraction are summarized as follows:

$(\bar{x} - s)$ to $(\bar{x} + s)$: 68%

$(\bar{x} - 2s)$ to $(\bar{x} + 2s)$: 95%

$(\bar{x} - 3s)$ to $(\bar{x} + 3s)$: 99.7%

The standard deviation changes with the number of data.   As the
number of data n tends to infinity, the standard deviation s tends
to a unique value σ.   The standard deviation s is called the sample standard deviation, and σ, the population standard deviation.
The standard deviation σ or s represents the repeatability of the
sensor.   Smaller standard deviation gives higher repeatability.
On the other hand, the bias $(x' - \bar{x})$ represents the accuracy
of the sensor, which depends on the calibration of the sensor as
mentioned above.

The sensor with high repeatability is applied to the case
where high precision is required.   The related term "resolution"
is defined as the measurement range divided by the minimum difference of the state variable which can be measured by the sensor.
Therefore, the sensor having the higher resolution can detect a
smaller change of the variable.

In relation to the third condition, we will discuss a sensor
signal x which is a function of several variables a_j, including the
state variable.   Then the following equation is derived from the
chain rule in mathematical differentiation [1]:

$$dx = \sum_{j=1}^{m} \frac{\partial x}{\partial a_j} \, da_j \qquad\qquad (2.2.2)$$

Now we assume that a_1 is the state variable to be measured. Then $|\partial x/\partial a_1|$ is the sensitivity of the sensor. If all remaining $\partial x/\partial a_j$ (j = 2, 3, ..., m) are zero, the sensor is not affected by variables other than the objective state variable, and the signal x is substantially equal to the state variable a_1.

If the signal x is affected by some other variables, the standard deviation σ is estimated by the following equation:

$$\sigma = \sqrt{\sum_{j=1}^{m} \left(\frac{\partial x}{\partial a_j} \sigma_j \right)^2} \qquad\qquad (2.2.3)$$

where σ_j is the corresponding standard deviation of a_j. Equation (2.2.3), called the law of propagation of error, is derived from Eq. (2.2.2). For example, state variable a_1 represents the powder flow rate in gas-solid two-phase flow, a_2 the temperature of the gas, a_3 the pressure, a_4 the moisture content of the solids, and so on. The relations $\partial x/\partial a_j$ (j = 2, 3, 4, ...) = 0 mean that the signal is determined solely by the powder flow rate a_1. If $\partial x/\partial a_2 \neq 0$, the signal is affected by the temperature of gas, and the temperature fluctuation causes an error in the sensor signal. Equation (2.2.3) is also useful for estimating the total error of a measuring system, which is composed of several units for signal transformation.

In relation to the fourth condition, the disturbance or noise is discussed here. In Eq. (2.2.2), a_j other than a_1 can act as the disturbance on the signal x. When the cause of disturbance is obscure, it is called noise. We usually assume a white noise, which is discussed below.

In relation to the fifth condition, the dynamic response of the sensor is discussed. The response is usually examined by the frequency response to a wavy signal. An input wave with higher frequency makes the amplitude of the output wave smaller. The frequency, which decreases the amplitude to $1/\sqrt{2}$ of the input wave, is called the cutoff frequency. If the cutoff frequency is low, the sensor cannot follow the fast change in the state variable, resulting in an error. This type of error is called dynamic error.

Now we get back to a discussion of white noise, denoted by $n(t)$. It has such a unique characteristic that there is no correlation between the noise $n(t)$ at time t and the noise $n(t + \tau)$ at time $t + \tau$. In other words, the autocorrelation $\phi(\tau)$ of the white noise is given by an impulse function $\delta(\tau)$:

$$\phi(\tau) = \delta(\tau)$$

$$\delta(\tau) = \begin{cases} 0 & (\tau \neq 0) \\ \infty & (\tau = 0) \end{cases} \qquad (2.2.4)$$

$$\int_{-\infty}^{\infty} \delta(\tau)\ d\tau = \text{const.}$$

The autocorrelation of an arbitrary signal $f(t)$ is defined by the following equation:

$$\phi(\tau) = \lim_{T \to \infty} \frac{1}{2T} \int_{-T}^{T} f(t)f(t + \tau)\ dt \qquad (2.2.5)$$

Equation (2.2.5) shows that the autocorrelation $\phi(\tau)$ is determined solely by τ. The parameter τ is called the delay-time parameter.

Power spectrum $\Phi(\omega)$, which is a function of angular frequency ω, is given by Wiener-Khinchin's relation:

$$\Phi(\omega) = \int_{-\infty}^{\infty} \phi(\tau)\ \exp(-j\omega\tau)\ d\tau \qquad (2.2.6)$$

where $j^2 = -1$. the quantity $\exp(-j\omega\tau)$ is composed of sinusoidal waves, as is clear from Euler's relation:

$$\exp(-j\omega\tau) = \cos \omega\tau - j \sin \omega\tau \qquad (2.2.7)$$

From Eqs. (2.2.4) and (2.2.6), the power spectrum for a white noise becomes

$$\Phi(\omega) = \int_{-\infty}^{\infty} \delta(\tau) \exp(-j\omega\tau) \, d\tau$$

$$= \text{const.} \qquad\qquad (2.2.8)$$

Equation (2.2.8) shows that the power of the white noise is constant in all frequency ranges.

Some types of sensors are based on the so-called correlation technique, and they utilize the following cross-correlation:

$$\phi_c(\tau) = \lim_{T\to\infty} \frac{1}{2T} \int_{-T}^{T} f_1(t)f_2(t + \tau) \, dt \qquad (2.2.9)$$

where f_1 and f_2 are any signals detected at a distance L.

It may also happen that with the passage of time, the measured value gradually deviates from the true value. The seventh condition relates to this problem. The error $x(t) - x'$, called the systematic error, is caused by either a failure of instruments or a change in the surrounding conditions. The characteristics of a sensor will be changed by chemical contamination or physical wear caused by frequent particle impact.

The final requirement is that the handling and maintenance have to be easy. In powder handling processes, fine particles may enter the instruments. These particles adhere to electric parts or mechanical bearings, resulting in an electrical or mechanical breakdown.

2.3 VARIOUS PHENOMENA APPLIED TO POWDER SENSORS

Measuring techniques are classified as direct measurement, such as microscopic analysis of particle size, or indirect measurement, such as sedimentation analysis of particle size distribution. The various methods described in this book belong to the latter group, that is, indirect measurement. The range of measurement is restricted by the principle of the method, especially for indirect measurement. Indirect measurement is also called inferential measurement, which requires calibration and, sometimes, a complicated calculation by use of a microcomputer.

Measurement in which physical contact with the material is not necessary is called noncontact measurement. Optical or ultrasonic techniques, for example, give noncontact measurement. Noncontact measurement does not disturb the process system. Whether the measurement is of contact or noncontact type is determined by the principles applied to the sensor.

Those principles of powder sensors are based on various physical or chemical phenomena, which are discussed briefly in this section. For the development of a new sensor, one of the phenomena will give a basis. It is also important to know the principle, because it is the main factor that determines a shortcoming of the sensor or the effective range of the measurements. Instruments such as a photodetector, voltmeter, or frequency counter may be different, depending on the sensor principle, and knowledge of the principle is an aid in the selection of additional instruments.

Here, we group the phenomena into the following 13 classes [2]:

1. Energy absorption (extinction)
2. Scattering (reflection, diffraction)
3. Electromagnetic phenomena
4. Electrostatic phenomena
5. Piezoelectric phenomena
6. Pressure drop
7. Doppler effect
8. Chemical reaction (thermodynamic phenomena)
9. Statistical phenomena
10. Mechanics
11. Holography
12. Acoustic phenomena
13. Nuclear magnetic resonance

Table 2.1 summarizes phenomena 1 and 2. Thermal energy will be absorbed by particles when they are heated. Particles may be solid or liquid, but their heat conductivity should be so high that the absorbed heat or temperature decrease of the surrounding gas can be detected by suitable methods. Then the flow rate of particles can be inferred.

Table 2.1 Various Phenomena as Principles of Powder Sensors (Phenomena 1 and 2)

Phenomena	Media	State variables	Note
1. Energy absorption (extinction)	Heat flux	Powder flow rate	Heat conductive particles
	Hot wire	Particle size	Mist (liquid particle)
	Radiation: β-ray	Concentration	Attenuation
	γ-ray	Level	
	Light extinction	Concentration	Dust
	Ultrasonic wave	Level	
	Microwave	Moisture content	
	Glow discharge	Particle size	Discharge current decrease
2. Scattering (reflection, diffraction)	Nuclear (neutron)	Moisture content	Detecting thermal neutron
	Infrared ray	Moisture content	
	Ultrasonic wave	Level, particle size	Reflection time
	Light scattering	Particle size	Dust, aerosol
	Laser diffraction	Particle size	
	Lidar	Concentration, particle size	Aerosol
	Photoreflection	Flow rate	Cross-sectional area of powder bed
	Electromagnetic pulse wave	Level	Transit time

Some types of radiation may go through a human body, but the energy of the radiation is partly absorbed by the body. Light, ultrasonic waves, and microwaves (electromagnetic waves) are also absorbed by particles. The energy absorption phenomenon is utilized in the measurement of various state variables, such as powder flow rate, particle size, particle concentration, moisture content, and level of powder bed. A typical example of phenomenon 2 (scattering, etc.) is light scattering by particles. The sky is blue because the small particles in the sky scatter blue light (light of short wavelength) in preference to red light. A reflection phenomenon, such as an echo, is utilized to measure the distance between the signal source to the object body where the signal is reflected, because the time delay multiplied by the speed of the signal equals twice the distance. Phenomenon 2 is utilized in the measurement of particle size, particle concentration, powder bed level, or moisture content of powder.

Table 2.2 lists phenomena 3, 4, and 5. Light absorption and scattering are also electromagnetic phenomena, because light is an electromagnetic wave. However, for convenience, we exclude them. They are included in phenomena 1 and 2.

The electric capacitance of a condenser depends on the materials contained between the two electrodes. Then the capacitance gives effective information for the measurement of powder concentration, powder bed level, and moisture content. Electric resistance or conductance is also useful in measuring the moisture content of particles. Electromagnetic induction, which is caused by the movement of electroconductive materials in a magnetic field, is applied in a slurry flow meter. Microwave resonance may provide a method for the measurement of powder flow rate.

Electrostatic phenomena include the static electrification of particles by contact or impact on the wall of particulate equipment, corona charging of fine particles, and inductive charging of particles in an electric field. These phenomena are utilized in the detection of powder flow, measurement of particle size, and number concentration of fine particles.

Piezoelectric phenomena are associated with a quartz oscillator (piezoelectric vibrator) or crystal dielectrics. The frequency of the quartz oscillator is changed by even a small dust deposit. On the other hand, if the crystal dielectrics (including quartz)

Table 2.2 Various Phenomena as Principles of Powder Sensors (Phenomena 3, 4, and 5)

Phenomena	Media	State variables	Note
3. Electromagnetic phenomena	Electric capacitance	Concentration	Holdup concentration
		Level	
		Moisture content	
	Electric resistance	Level	RC circuit
		Moisture content	
	Electromagnetic induction	Velocity	Slurry
	Microwave resonance	Concentration, velocity (flow rate)	
4. Electrostatic phenomena	Electric charge (electric current)	Powder flow rate	Static electrification
		Particle size	Induction charging, mist
			Corona charging, dust
		Concentration	Induction charging, mist
			Corona charging, dust
5. Piezoelectric phenomena	Piezoelectric vibration	Concentration	Dust
	Electric pulse	Particle size	Impulsive force
		Flow rate	Local flow rate

are mechanically strained, an electric pulse will be produced.
The former is utilized in dust concentration measurement, and
the latter is utilized in the measurement of particle momentum
(mass × velocity) and local number concentration.

Phenomena 6, 7, 8, and 9 are listed in Table 2.3. Flow
through a powder bed induces a pressure drop which depends
on several parameters, including particle size. Therefore, particle size is estimated by the pressure drop. The pressure drop
for gas-solid two-phase flow is also a function of particle concentration. Further, the backpressure of an air jet kept at a
constant flow rate increases when the jet impinges against the
powder flow or powder bed. These phenomena are utilized to
detect the level of the powder bed or the powder flow rate.

The Doppler effect is a phenomenon in which the frequency
of sound, observed at a fixed point, changes with the velocity
of the sound source. The same effect is observed for the movement of a medium through which the sound wave passes. A
laser can be used instead of sound, as long as the scattering
particles are suspended in the medium. The phenomenon can be
applied to the velocity measurement of fluid or particles. These
methods do not require physical contact with the objects.

Chemical reactions or thermodynamical phenomena may be applied to powder sensors, but applications are not yet common.
For an off-line measurement of moisture content, an ethanol solution of cobalt chloride is utilized because the color of the solution
changes from light blue to pink through a reaction with moisture.
The heat of immersion between the powder and some types of
solvent is utilized in off-line measurement of particle size because
the heat is directly proportional to the specific surface area of
powder.

Condensation of a supersaturated vapor (water, alcohol, etc.)
is utilized in the pretreatment of an aerosol for the concentration
measurement of very fine particles (less than $0.3 \ \mu m$). The particles act as the nuclei for the condensation, and they will grow
large enough to be detected by light extinction or scattering.
A related physical phenomenon is known as the Kelvin effect.

Statistical phenomena—group 9 in the table—include the turbulent phenomena in a flow and the acoustic, electric, or electromagnetic noise caused by random movement of particles suspended
in a flowing fluid. The Brownian motion of particles is also a

Table 2.3 Various Phenomena as Principles of Powder Sensors (Phenomena 6, 7, 8, and 9)

Phenomena	Media	State variables	Note
6. Pressure drop	Air or gas pressure	Level	Back pressure
		Powder flow rate	Pressure loss, back pressure
		Particle size	Permeability
7. Doppler effect	Ultrasonic wave	Fluid velocity	Slurry
	Laser	Fluid velocity	
		Particle velocity	Low concentration
8. Chemical reaction	Cobalt chloride	Moisture content	
	Heat of immersion	Particle size	
	Condensation	Particle size, concentration	Aerosol
9. Statistical phenomena	Correlation	Concentration, velocity, (flow rate)	Noise
	Spectrum	Flow rate	Acoustic
	Eddy scale	Particle size	
	Diffusion	Particle size	Aerosol, fine particles

statistical phenomenon.   These statistical phenomena are utilized
in measurements by taking an autocorrelation [see Eq. (2.2.5)]
or a cross-correlation [see Eq. (2.2.9)] of the signals.   Despite
the random movement of fluid or particles, there is a correlation
between the two signals detected at a short distance.   Then the
velocity of particles can be measured by the cross-correlation.
Such a measurement is also known as the correlation technique,
mentioned in the preceding section.

The last grouping in Table 2.4 are phenomena 10, 11, 12,
and 13.   The mechanics of a particle or particulate materials give
the fundamentals of measurements on the level of powder bed,
powder flow rate, particle size, and so on.   The mass of the particulate materials contained in a storage vessel is continuously
measured by a load cell, which transforms the weight (= mass ×
gravity) into an electrical signal through, for example, an electromagnetic phenomenon.   Holography is three-dimensional photog-
raphy utilizing a pulsed laser beam.   It is possible with holography
to determine the particle velocity, size, local concentration, and
trajectory.   The analysis is complicated, however, and is not practical for on-line measurement.   Acoustic emission is caused by
impact or friction between particles, or between particles and the
wall of a piece of powder handling equipment.   Spectrum analysis
of the acoustic emission may give a method for measurement of
powder flow rate, particle size, and the movement of particles in
a vessel.   It is also possible to count the number of particles by
individually tracking the impact sound of a single particle.   Nuclear magnetic resonance (NMR) is one of the electromagnetic
phenomena, but is not yet in common use in on-line measurements
except for moisture measurement.

The phenomena described above are applied to the various
types of powder sensors as their measurement principles.   Some
of them are put to practical use, and some are under development.
Each of the phenomena listed here has a specific scientific field,
such as mechanics, optical science, or electromagnetics.   More de-
tailed descriptions of the phenomena are given in subsequent
chapters in relation to each method described.

The following phenomena, which are not described above, may
also be applied to some types of sensors: Wiedemann effect, magne-
toresistance effect, thermomagnetic effect (Righi-Leduc effect,

Table 2.4 Various Phenomena as Principles of Powder Sensors (Phenomena 10, 11, 12, and 13)

Phenomena	Media	State variables	Note
10. Mechanics	Gravity	Level Powder flow rate	Level meters Belt scale, etc.
	Coriolis force	Powder flow rate slurry flow rate	
	Centrifugal force	Particle size, etc.	
11. Holography	Pulse laser	Velocity (local) concentration, trajectory	Not yet practical, only for research work
12. Acoustic phenomena	Acoustic emission	Velocity, particle size, particle number	Suspension
13. Nuclear magnetic resonance	Electromagnetic wave Larmor precession	Moisture content Velocity	Slurry

Nernst effect), photoelectromagnetic effect, and aeolian tone (von Kármán's vortex street).

REFERENCES

1. T. M. Apostol, Mathematical Analysis, Addison-Wesley Publishers, Ltd., London, Chap. 6 (1957).
2. H. Masuda and K. Iinoya, J. Res. Assoc. Powder Technol. Jpn., 14, 228 (1977).

3

Sampling of Particles

Sampling of particles will sometimes be required in order to ex-
amine the physical condition of powder handling processes. In
Section 3.1 we describe the sampling from powder flow or from
a powder bed, and in Section 3.2 the sampling from gas-solid
suspension flow.

3.1 POWDER SAMPLING

Powder sampling is an important procedure not only for the ex-
amination of powder handling processes but also for determination
of the quality of powder products. For example, chemical composition, particle size distribution, moisture content, and so on,
will be measured by use of a small amount of the final products.
Commercial transactions concerning the products will be made
based on the results of the measurements. The sampled powder
should therefore adequately represent the powder quality as a
whole.

3.1.1 Sampling Equipment

When powder is poured into a heap, size segregation will occur. Fine particles tend to remain at the center of the heap, and coarse particles tend to collect at the periphery. Even when the powder is transported slowly on a belt conveyor, coarse particles tend to collect on the surface and both sides of the powder bed because of the mechanical vibration of the conveyor. Size segregation occurs more easily for powders with higher flowability. Sampling from a powder bed or powder heap is therefore not preferred.

The following two rules are called the golden rules of sampling [1]:

1. A powder should be sampled when it is in motion.
2. The entire stream of powder should be taken for many short increments of time.

The first rule recommends sampling from flowing powder when it is discharged from a belt conveyor or when it is poured from a storage vessel into another vessel. Even in these cases, size segregation may occur. Therefore, the entire stream of powder should be sampled by traversing the stream for many short increments of time in preference to part of the stream being taken for the entire period. If the sampling speed v (m/s) is low, the mass of the sampled powder (called an increment) may be too great. The sampled mass w (kg) will be calculated by the following equation (see also Fig. 3.1):

$$w = \frac{L}{v} \frac{W}{L} b = \frac{Wb}{v} \qquad\qquad (3.1.1)$$

where L (m) is the width of the powder stream, W (kg/s) the powder flow rate, and b (m) the width of the sampler's mouth (sampling width or cutter width).

Equation (3.1.1) shows that the sampled mass is inversely proportional to the sampling speed and therefore can be reduced by increasing the sampling speed. It is assumed in Eq. (3.1.1) that the powder flows uniformly over the flow width L and that the sampling width b is much larger than the particle size. If the width b is 20 times larger than the maximum particle diameter, the effective sampling width is larger than 0.95b.

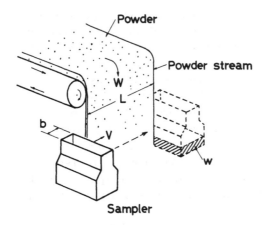

Figure 3.1   Powder sampling.

A sampler that follows the foregoing rules is called a full-stream sampler.   One full-stream sampler, the cutter sampler, is shown schematically in Fig. 3.2. There are several types of full-stream samplers, including the car-type, belt, and rotary samplers [2].   Sampling width (cutter width) and sampling speed are adjustable in these samplers.   Therefore, it is possible to take the desired mass of powder according to Eq. (3.1.1).   Random sampling is also possible by randomly changing the starting time of the sampler.

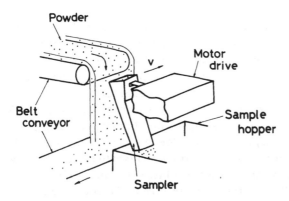

Figure 3.2   Full-stream sampler (cutter sampler).

As discussed above, the full-stream sampler is preferred. Also used, however, are snap samplers and slide spoon samplers. These samplers take a small amount of powder from several positions of a powder stream or powder bed [2]. A specially designed spoon is utilized in such a sampler. The mass of the sampled powder (increment) is much less than that obtained by the full-stream sampler, and the final sample obtained by gathering all increments is easily treated in the laboratory measurement. The sample is, however, easily affected by size segregation, especially when a sample is taken from a powder bed formed on a belt conveyor. It is therefore required that the spoon should dip up the powder from as deep a location as possible, to avoid the error caused by the size segregation.

A scoop will also be used to take a sample from powder loaded on a truck. In this case it is recommended that the sampling be done after digging 30 cm from the surface, because the surface region is usually affected by segregation [3]. Sampling spears can be used to take samples from deep regions of the powder bed. A sampling device of air-suction type, which is similar to pneumatic ship unloaders, is also useful to take samples vertically from a powder bed. The sampling probe consists of two co-cylindrical tubes, and suction air comes into the annular gap between the two tubes from the basal part of the sampler. Particles are mixed with air at the sampler nose and the resulting gas-solid suspension is sucked into a gas-solid separator such as a cyclone or a bag filter [1]. Sampling positions in these cases are determined so as to divide the powder bed into equal-area regions.

When a continuous sampling from a storage vessel is required, a screw conveyor will be utilized. A table feeder with multi-scrapers, developed in our laboratory, is also useful in taking small samples during continuous feeding of powder to a process [4]. Particles passing through the table feeder do not suffer from breakage, whereas the screw conveyor causes particle breakage. A special constant-volume sampler or slide-valve sampler may be applied in the sampling of powder that is sliding on an inclined chute. These samplers are suddenly inserted into the powder stream and therefore may cause problems in the mechanical parts of samplers. Also, if the powder stream is only partially sampled, size segregation cannot be avoided.

The powder gathered from all increments is called the gross sample. The mass of the gross sample is usually too great to use as a sample for laboratory measurement of chemical composition, particle size distribution, and so on. It is also necessary to reduce the mass of the sample before sending it to the laboratory. The procedure used to reduce the sample mass is called sample dividing.

Sample dividing should be done without segregation. The principles involved here are the same as those already discussed when considering collection of the gross sample. That is, the sample should be taken when the powder is in motion and the entire stream of powder should be taken for many short increments of time rather than taking a part of the stream continuously. The sample divider shown schematically in Fig. 3.3, called a spinning riffler, may satisfy the golden rules of sampling. The performance quality of the sample divider depends on the rotational speed. Higher speed (as high as 300 rpm) gives better results.

There is a different type of divider in which the feeding device rotates over several receivers and feeds particles into them. A sample divider of this type is called a mechanical distributor or sample splitter. Similar but simpler sample dividers are the oscillating hopper sample divider and the oscillating paddle sample divider [1]. The oscillating hopper sample divider has a feed hopper which is pivoted about a horizontal axis so as to swing like a pendulum. The hopper feeds the sample powder

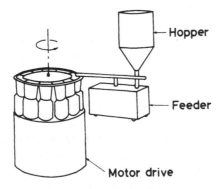

Figure 3.3 Spinning riffler.

alternately into two receivers placed under the hopper outlet.
The contents of one receiver are retained, and the mass of sample is thus halved at each step. In the oscillating paddle sample
divider, the feed hopper is fixed vertically. There is an oscillating paddle just below the feed hopper and it divides the powder stream right and left. These two sample dividers are not as
effective in reducing the sample mass as are the spinning riffler
and mechanical distributor.

All of the sample dividers described above have moving parts,
which will cause trouble in a dusty atmosphere. On the other
hand, a sample divider called a chute riffler or chute splitter has
no moving parts. It has two series of chutes directed right and
left, one after the other. The number of chutes is the same on
the right-hand and left-hand side. Powder poured onto the
chute riffler is divided by these two series of chutes. The sample dividing by the chute riffler is affected by particle segregation, although the effect is reduced as the number of chutes is
increased. The segregation effect will be diminished by the following procedure.

Repeat the sample dividing three times, as shown in Fig. 3.4,
taking note of both the right-hand-side and the left-hand-side
samples. Then we get eight samples after the third step of dividing. From these eight samples, we get four samples by adding two of them as shown in the figure. Each of these four samples contains particles that are divided three times to the right
and three times to the left. The effect of the size segregation is
thus fairly reduced in these samples. The procedure described
above is called Carpenter's method [5].

The chute riffler needs repeated division to reduce the sample
size (mass). A sample divider called a table sampler or multicone
divider can, however, greatly reduce the sample size by only one
trial. The table sampler has a series of holes and prisms on an
inclined table. The powder to be divided is fed to the top of the
sampler (divider). Prisms placed in the path of the powder
stream break it into fractions. Some particles fall through the
holes and are discarded; the powder remaining on the table passes
on to the next row of prisms and holes, more powder is removed,
and so on. The powder reaching the bottom of the table is the
final sample. Sample dividers similar to the table sampler but
with no holes are called the 1/16 divider, 1/32 divider, and so on,
according to the number of final samples obtained.

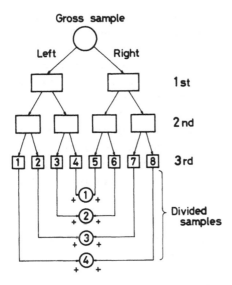

Figure 3.4 Carpenter's method for sample dividing.

These sample dividers need uniform distribution of the initial feed and complete mixing after each separation, but it is difficult to achieve such a condition. Therefore, the sample obtained by these dividers will suffer from fluctuations in the sample size and the particle size distribution.

Some sample dividers are applied directly in the process [6]. They are called the slotted-belt sample divider, rotary sample divider, and a chain bucket sample divider. The divided samples may, however, need further reduction in sample size when they are tested in the laboratory. Other samplers, such as a chain bucket sampler, swing arm sampler, and rotary sampler, are also described in the literature [6]. When a hazardous powder must be sampled, the sampling device should protect the worker from dangerous conditions or exposure at all stages of the sampling cycle [7].

3.1.2 Analysis of Samples

Various types of data will be obtained by analyzing the sampled powder. The mass median diameter (MMD) of the sample, for example, can be obtained through the particle size measurement.

The value of MMD is, however, dependent on the sample. Different values will be obtained for different samples. The whole set of these values is called the population of the data on the mass median diameter of the powder. The value x_1 is only a first sample extracted from the population with a certain probability. Such a variable, like the MMD, is called the random variable. The random variable x is characterized by a so-called probability density function or frequency distribution f(x). The probability that the sampled value is found in the range x to x + dx will be given by f(x) dx. The following equation is very useful for many types of random variables that appear in natural phenomena or in industrial processes:

$$f(x) \ dx = \frac{1}{\sqrt{2\pi}\ \sigma} \ \exp\left[- \frac{(x - \mu)^2}{2\sigma^2} \right] \ dx \qquad (3.1.2)$$

where μ is called the population mean and σ^2 is the population variance.

The function f(x) is called the normal distribution or Gaussian distribution. If we introduce a new variable $z = (x - \mu)/\sigma$, Eq. (3.1.2) can be rewritten as follows:

$$f(z) \ dz = \frac{1}{\sqrt{2\pi}} \ \exp\left(- \frac{z^2}{2} \right) \ dz \qquad (3.1.3)$$

The function f(z) is the standard normal distribution, and the integral from 0 to z for various values of z will be found in a table [8]. The integral from $z = -2(x = \mu - 2\sigma)$ to $z = 2(x = \mu + 2\sigma)$ is, for example, about 0.95, as described briefly in Section 2.2.

The population mean μ and the population variance σ^2 are unknown parameters of Eq. (3.1.2). These parameters should be estimated by the following sample mean $\bar{x}$ and sample variance s^2:

$$\bar{x} = \frac{1}{n} \sum_{j=1}^{m} x_j f_j \qquad (3.1.4)$$

$$s^2 = \frac{1}{n-1} \sum_{j=1}^{m} (x_j - \bar{x})^2 f_j \qquad (3.1.5a)$$

$$= \frac{1}{n-1} \sum_{j=1}^{m} \left[x_j^2 f_j - \frac{1}{n} \left(\sum_{j=1}^{m} x_j f_j \right)^2 \right] \qquad (3.1.5b)$$

where x_j is the representative value of the jth class, f_j the frequency of x_j (j = 1, ..., m), and n the sample size (n = $\Sigma_{j=1}^{m} f_j$).

The fact that the denominator of Eq. (3.1.5a) is not n but n − 1 is to make the variance unbiased. If the sample size n is large enough, it can be replaced by n. The square root of the sample variance is called the sample standard deviation.

The confidence interval of the population mean with confidence level 1 − p is given by the following equation:

$$\bar{x} - t(\nu,p) \frac{s}{\sqrt{n}} < \mu < \bar{x} + t(\nu,p) \frac{s}{\sqrt{n}} \qquad (3.1.6)$$

The values of $t(\nu,p)$ will be found in a table [8]. Some of them are listed in Table 3.1. The definition of t is as follows:

$$t = \frac{\bar{x} - \mu}{s} \sqrt{n} \qquad (3.1.7)$$

Table 3.1 Student's t Distribution $t(\nu,p)$

ν/p:	0.25	0.05	0.01
3	1.4226	3.1825	5.8409
4	1.3444	2.7764	4.6041
5	1.3009	2.5706	4.0321
6	1.2733	2.4469	3.7074
7	1.2543	2.3646	3.4995
8	1.2403	2.3060	3.3554
9	1.2297	2.2622	3.2498

This is also a random variable and it follows Student's t distribution with degree of freedom $\nu = n - 1$. If the population mean μ is obtained, $\bar{d} = \bar{x} - \mu$ is called the bias.

On the other hand, the confidence interval of the population variance with confidence level $1 - p$ is given by the following equation:

$$\frac{ns^2}{\chi^2(\nu,p)} < \sigma^2 < \frac{ns^2}{\chi^2(\nu, 1 - p)} \tag{3.1.8}$$

The definition of χ^2 is as follows:

$$\chi^2 = \frac{ns^2}{\sigma^2} \tag{3.1.9}$$

This is also a random variable which follows the χ^2 distribution.

The precision of the sampling is usually represented by 2σ [3]. The standard deviation is called the average error and $(2/3)\sigma$, the probable error. The quantity CV, the coefficient of variation, is more often utilized in representing the precision of the sampling.

$$CV = \frac{\sigma}{\mu} \times 100 \quad \% \tag{3.1.10}$$

On the other hand, the accuracy of the sampling is represented by the bias $\bar{d}$. Now we consider, as an example, whether or not a snap sampling is accurate by assuming that the full-stream sampling is accurate. Let the data obtained by the full-stream sampling be represented by x_i and the data by the snap sampling x_i'. The procedure is as follows:

1. Calculate $d_i = x_i - x_i'$ for $i = 1, 2, \ldots, n$.
2. Calculate the variance of d [by the use of an equation similar to Eq. (3.1.5)].
3. Calculate t by the use of Eq. (3.1.7) with $\bar{d} = \bar{x} - \mu$.
4. Find the value of $t(\nu, 0.05)$ from Table 3.1.
5. Compare t with $t(\nu, 0.05)$. If $|t| > t(\nu, 0.05)$, the snap sampling has a significant bias with confidence level of 95%. The bias is given by $\bar{d}$.

Example 3.1.1   The sampling was carried out with a full-stream
sampler and a snap sampler one after the other.   The mass median diameters x_i and x_i' (i = 1 to 10) were obtained.   The mean
difference of these mass median diameters was 20 μm and the
standard deviation was 10 μm.   Is the snap sampling adequate
as an alternative to full-stream sampling?   Discuss the problem
assuming a confidence level of 95%.

   Solution: From Eq. (3.1.7), the t value for the mean difference of the mass median diameters is (t = 20 × $\sqrt{10}$/10 =)6.32.
As the confidence level is 95%, p = 0.05.   Then t(9, 0.05) in
Table 3.1 is 2.26.   As the value of t calculated by Eq. (3.1.7)
is larger than t(9, 0.05), the snap sampling is not adequate as
an alternative of the full-stream sampling.   The bias of the snap
sampling compared with the full-stream sampling is 20 μm.

   There are many types of error affecting the final data.   The
mass median diameters in Example 3.1.1 are affected by the sampling error, sample dividing error, and the size measurement
error.   The following equation is applied in an analysis of the
total error, as discussed in Section 2.2.

$$\sigma_t = \sqrt{\sum_{j=1}^{m} \left(\frac{\partial x}{\partial a_j} \sigma_j \right)^2} \qquad (3.1.11)$$

where a_j represents some type of error and σ_j is the standard
deviation of a_j.   As the sampling, the sample dividing, and the
size measurement are independent of each other, the standard
deviation of the final data will be estimated by the square root
of the summation of the unbiased variance of each operation:

$$\sigma_t = \sqrt{\sum_{j=1}^{k} \sigma_j^2} \qquad (3.1.12)$$

   The mean particle diameters discussed in Section 1.4 are
usually determined through size measurement of the sampled particles.   The mean particle diameter utilized in many particulate
processes is given by the following equation, as discussed in
Section 1.4:

$$\bar{D}_p = \left[\int D_p^m f(D_p) \, dD_p \right]^{1/m} \tag{3.1.13a}$$

$$= \exp\left(\mu^{(\beta)} + \frac{1}{2} m\sigma^2 \right) \tag{3.1.13b}$$

where m is the exponent representing the particulate process proportional to D_p^m and σ is the standard deviation of the particle size distribution [Eq. (1.3.9)]. Equation (3.1.13a) is for general use and Eq. (3.1.13b) is only for log-normal size distribution.

The value of the mean particle diameter depends on the number of particles sampled. If the number of particles is not high enough, the mean particle diameter deviates from the corresponding true value even if the sampling and sample dividing are carried out perfectly. The following equations are used to calculate the number of particles, n^*, required in the determination of the mean particle diameter within the relative error of δ with probability p [9]:

$$n^* = \frac{\omega}{\delta^2} \tag{3.1.14a}$$

$$\omega = u^2 m^2 \sigma^2 (2c^2\sigma^2 + 1) \tag{3.1.14b}$$

$$c = \beta + \frac{1}{2} m \tag{3.1.14c}$$

where u is the parameter that depends on the preassigned probability p. Table 3.2 gives the value of u as a function of the probability. Equation (3.1.14) can also be utilized in estimating the error δ when the number of particles used in the measurement is given.

Table 3.2 u Values in Eq. (3.1.14)

p (unitless)	0.5	0.75	0.9	0.95	0.975	0.99	0.995
u (unitless)	0.67	1.15	1.64	1.96	2.24	2.57	2.81

It is usually not possible to get enough particles in the on-
line measurement of process variables relating particle size, because the number of particles detected by a sensor depends on
the process conditions. Therefore, the mean particle size detected will fluctuate according to Eqs. (3.1.14), even if the sampling is perfect in the usual sense. If the number of particles
sampled is too small, the detected value will have a bias given by
the following equation [9]:

$$d = \bar{D}_p - \bar{D}_p' = \bar{D}_p \left[1 - \exp\left(\frac{c\sigma^2}{n^*}\right) \right] \qquad (3.1.15)$$

where $\bar{D}_p'$ is the true value of the mean particle diameter.

Example 3.1.2   The population of particles follows the log-normal
size distribution on the count basis. The size distribution has
a 50% diameter of 50 μm and a 16% oversize diameter of 80 μm.
Now we want to estimate the mean volume diameter of these particles within a relative error of ±5% with 95% probability. How
many particles should be measured?

Solution: From Eq. (1.3.11), σ = 0.47. In this case β = 0
(count basis) and m = 3 (mean volume diameter). Therefore, Eq.
(3.1.14c) gives c = 1.5. The u value will be obtained from Table
3.2 as 1.96. Substituting these values into Eq. (3.1.14b) gives

$$\omega = (1.96)^2(3)^2(0.47)^2[2(1.5)^2(0.47)^2 + 1] = 15.2$$

As δ = 0.05 (5% relative error), Eq. (3.1.14a) gives

$$n^* = \frac{15.2}{(0.05)^2} = 6080$$

3.2   SAMPLING OF GAS-SOLID SUSPENSIONS

3.2.1   Isokinetic Sampling

When the particles are suspended in a flowing gas, a suction-
type of sampling probe will be utilized, as shown in Fig. 3.5.

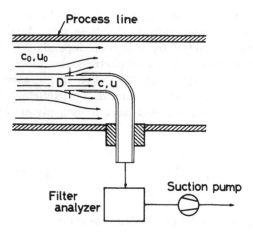

Figure 3.5   Sampling from suspension flow.

The sampled particles are transported through the tube and will
be analyzed by the use of several types of instruments.   The
velocity of the sampled suspension flow at the inlet of the sampling probe should be equal to the velocity of the main flow.
The sampling that satisfies the condition is called isokinetic sampling.   If the sampling is not isokinetic, the sampling suffers
from so-called anisokinetic sampling error, which is caused primarily by particle inertia.   There are many empirical or semiempirical equations for estimating the anisokinetic error [10].   Some
of them are listed in Table 3.3.   Numerical calculations have been
carried out and some of the results are listed in Table 3.4.   The
simplest but most satisfactory estimations are done by the use of
the following empirical equation [12] when the velocity ratio
u_0/u is between 0.5 and 1.5:

$$\frac{c}{c_0} = \frac{u_0}{u} \frac{4\psi}{4\psi + 1} + \frac{1}{4\psi + 1} \tag{3.2.1}$$

where c_0 is the true concentration in the main flow, c the measured concentration, u_0 the main flow velocity, u the sampling
velocity, and ψ the inertia parameter.

Table 3.3 Empirical or Semiempirical Equations for Anisokinetic Sampling Error[a]

$$\frac{c}{c_0} = \frac{u_0}{u}\left[1 + F\left(\sqrt{\frac{u}{u_0}} - 1\right)\right]^2 \; ; \; F = \text{function of } \psi$$

Watson (1954)

$$\frac{c}{c_0} = 1 + \alpha\left(\frac{u_0}{u} - 1\right) \; ; \; \alpha = \frac{\psi D}{L}\left[1 - \exp\left(-\frac{L}{\psi D}\right)\right] \; ; \; L = 6 - 1.6D$$

Badzioch (1959)

$$\frac{c}{c_0} = \frac{u_0}{u} + K\left(1 - \frac{u_0}{u}\right) \; ; \; K = [1 + \exp(1.04 + 2.06 \log \psi)]^{-1}$$

Zenker (1971)

$$\frac{c}{c_0} = 1 + \frac{2[1 + 0.31(u/u_0)]\psi}{1 + 2[1 + 0.31(u/u_0)]\psi}\left(\frac{u_0}{u} - 1\right) \; ; \; \begin{array}{l} 0.18 < (u/u_0) < 6.0 \\ 0.18 < \psi < 2.03 \end{array}$$

Belyaev-Levin (1974)

[a]See also Eq. (3.2.1) and Table 3.4.

Table 3.4   Anisokinetic Sampling Error (Numerical Calculation)

u_0/u	2ψ	c/c_0	
		Numerical calculation	Eq. (3.2.1)
0.025	0.035	0.959	0.936
	0.08	0.914	0.865
	0.2	0.789	0.721
	0.5	0.583	0.513
	1	0.401	0.349
	2	0.237	0.220
	4	0.133	0.133
	10	0.069	0.071
0.05	0.035	0.939	0.938
	0.08	0.882	0.869
	0.2	0.739	0.728
	0.5	0.525	0.525
	1	0.354	0.366
	2	0.228	0.240
	4	0.147	0.155
	10	0.093	0.095
0.1	0.035	0.916	0.941
	0.08	0.854	0.876
	0.2	0.706	0.743
	0.5	0.491	0.550
	1	0.361	0.399
	2	0.260	0.280
	4	0.192	0.199
	10	0.141	0.143
0.2	0.035	0.915	0.947
	0.08	0.852	0.889
·	0.2	0.675	0.771
	0.5	0.532	0.600
	1	0.429	0.466
	2	0.343	0.360
	4	0.283	0.288
	10	0.237	0.238
0.35	0.035	0.863	0.957
	0.08	0.815	0.910
	0.2	0.733	0.814
	0.5	0.623	0.675

Table 3.4   (Continued)

u_0/u	2ψ	c/c_0	
		Numerical calculation	Eq. (3.2.1)
0.35 (cont.)	1	0.539	0.566
	2	0.468	0.480
	4	0.418	0.422
	10	0.380	0.381
0.7	0.035	0.943	0.980
	0.08	0.923	0.958
	0.2	0.881	0.914
	0.5	0.828	0.850
	1	0.790	0.799
	2	0.754	0.760
	4	0.729	0.733
	10	0.711	0.714
1.5	0.035	1.077	1.033
	0.08	1.109	1.069
	0.2	1.174	1.143
	0.5	1.267	1.250
	1	1.337	1.332
	2	1.400	1.400
	4	1.436	1.443
	10	1.468	1.476
2.0	0.035	1.172	1.065
	0.08	1.230	1.138
	0.2	1.351	1.285
	0.5	1.531	1.500
	1	1.676	1.665
	2	1.793	1.800
	4	1.877	1.887
	10	1.943	1.952
2.5	0.035	1.269	1.098
	0.08	1.348	1.207
	0.2	1.526	1.428
	0.5	1.793	1.750
	1	2.011	1.998
	2	2.197	2.200
	4	2.331	2.331
	10	2.413	2.428

Source: From Ref. 21.

The inertia parameter is defined by the following equation, discussed in Section 4.4:*

$$\psi = \frac{\rho_p D_p^2 u_0}{18 \mu D}$$

(3.2.2)

where D is the inlet diameter of the sampling probe.

If the inertia parameter is assumed to be zero, Eq. (3.2.1) is simplified as follows:

$$\frac{c}{c_0} = 1$$

(3.2.3)

Therefore, the data do not suffer the anisokinetic sampling error. On the other hand, if the inertia parameter is much larger than 1/4 and/or $\psi \gg [(u/u_0) - 1]/4$, Eq. (3.2.1) is simplified as follows:

$$\frac{c}{c_0} = \frac{u_0}{u}$$

(3.2.4)

Equation (3.2.4) means that the particle mass flow rate $(\pi/4)D^2 cu$ is not affected by the sampling velocity. In this case the measured concentration is not equal to that of the main flow. It is, however, possible to estimate the true concentration by the use of Eq. (3.2.4). Therefore, isokinetic sampling is not necessary in these two limiting cases.

If the inertia parameter takes an intermediate value, the amount of particles sampled will depend on the parameter. The measured concentration is always higher than the true concentration when the sampling velocity is lower than the main flow velocity. If the sampling velocity is higher than the main flow velocity,

---

*If the particle diameter is smaller than 10 μm, Cunningham's slip correction factor C_m should be multiplied. The slip correction factor is given by Eq. (3.2.12).

the measured concentration is always lower than the true concentration. Further, the sampling efficiency (= cu/c_0u_0) is dependent
on the particle size, because the inertia parameter is directly proportional to the square of the particle diameter. Therefore, the
size distribution of the sampled particles differs from that of particles suspended in the main flow. It is possible but complicated
to correct the measured concentration by applying Eq. (3.2.1)
to particles of different sizes. The deposition of particles in the
sampling tube should also be taken into consideration in the correction. As data correction is very difficult, isokinetic sampling
is, in general, preferable.

     Isokinetic sampling needs some consideration as to the sam-
pling head. A pressure control so as to equate the static pressure at the inside and outside walls of the sampling probe is not
sufficient for isokinetic sampling. The reason is that the additional pressure drop in the sampling probe should be compensated
in the control [13,14]. If the sampled particles are collected by
a filter, pressure control becomes difficult because the pressure
drop across the filter gradually increases. A special regulator
system is developed to avoid this problem [15]. Although the
original regulator system is mechanically constructed, only the
principle of the system is described here. Figure 3.6 shows a
diagram of the system. The system contains two air pumps, and
sampled suspension is diluted by adding clean air. Now the iso-

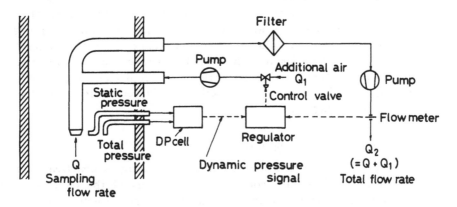

Figure 3.6 Flow-regulating systems of isokinetic sampling.

kinetic sampling flow rate is assumed to be Q and the additional
air flow rate Q_1. As the pressure drop across the filter increases, the flow rate Q_2 (= Q + Q_1) through the filter decreases.
Then the regulator system works and the clean airflow rate Q_1
is decreased so that the sampling flow rate Q remains unchanged.

The sampling probe should be aligned with the main flow direction. If the sampling probe is set with some angles to the
flow, sampling error cannot be avoided even if the sampling is
isokinetic [16–18]. The error also depends on the inertia parameter as in the case of aligned sampling. Therefore, numerical
correction of error caused by misalignment of the sampling probe
may be very complicated. If the inertia parameter is very small
($<$ 10^{-3}), the error caused by both misalignment and anisokinetic
sampling will be negligibly small. As for the tip of the sampling
probe, a sharper tip is recommended because the sample obtained
by a round tip will be affected by particle rebound on the tip.
A blunt tip should be avoided. Further, the sampling position in
the processes should be selected carefully so as to avoid the effect of flow distortion and to take representative samples [19].
Sampling from high-temperature suspension flow should be done
carefully with thermal insulation. When the hot gas is cooled in
the sampling line, vapor components will condense on the inside
wall, resulting in erroneous measurement of the sampling velocity
and sampled mass.

Electrostatic effects on the isokinetic sampling should also be
taken into consideration [20]. The measured concentration is
lower or higher than the true concentration, depending on the
polarity of the particles and the sampling probe. The sampling
error has a maximum value in a certain range of particle diameter
(1 to 10 μm). The smaller the sampling probe, the greater the
electrostatic effects. If it is difficult to avoid electrostatic effects,
the high-speed sampling discussed below is recommended instead
of isokinetic sampling.

3.2.2 High-Speed Sampling

There are some cases where isokinetic sampling is practically impossible. For example, continuous monitoring of the particle concentration may be required in a large-scale gas reactor, where
the gas is almost stationary. Sampling from an industrial clean

room may be included in these cases. Even if the gas flows slowly, the amount of particles obtained by the isokinetic sampling will be too small to be analyzed. In these cases it is better to carry out a high-speed sampling [21].

Figure 3.7 shows the concentration ratio c/c_0 as a function of the inertia parameter and the gravitational parameter. The inertia parameter is similar to that defined by Eq. (3.2.2), but the sampling velocity u is substituted instead of u_0. The gravitational parameter G is the terminal settling velocity (Stokes law) u_t divided by the sampling velocity u:

$$G = \frac{\rho_p D_p^2 g}{18\,\mu u} \tag{3.2.5}$$

If particles are assumed to be inertialess, the concentration ratio is given by the following equation:

$$\lim_{\psi \to 0} \frac{c}{c_0} = \begin{cases} 1 + G & \text{for upward sampling} \quad (3.2.6a) \\ 1 - G & \text{for downward sampling} \quad (3.2.6b) \end{cases}$$

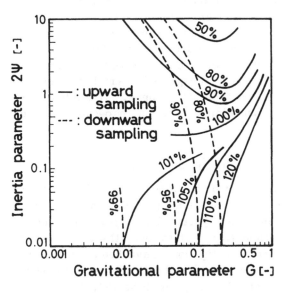

Figure 3.7   Concentration ratio c/c_0 for high-speed sampling.

These equations are obtained easily by taking the particle mass balance. The solid lines in Fig. 3.6 give the concentration ratio for upward sampling and the dashed lines that for downward sampling. When 2ψ is 0.01, both the 101% line for upward sampling and the 99% line for downward sampling assume a G value of 0.01, as shown by Eq. (3.2.6). Equation (3.2.6) holds for other cases when the inertia parameter 2ψ is smaller than 0.01.

Further, the following equations will be obtained:

$$2\psi G = \frac{2(\rho_p D_p^2)^2 \underline{g}}{(18\mu)^2 D} \qquad (3.2.7)$$

$$\frac{2\psi}{G} = \frac{2u^2}{D\underline{g}} \qquad (3.2.8)$$

From Eq. (3.2.7), the product $2\psi G$ is independent of the sampling velocity u. Therefore, if the sampling velocity is increased or decreased, the sampling condition changes along the line of slope -1 in the figure. On the other hand, Eq. (3.2.8) shows that the ratio $2\psi/G$ is independent of particle size. The concentration ratio for a different size is therefore given as the intersecting point between the line of slope 1 in the figure and the line of constant concentration ratio. The particle size can be determined by use of the definition of either the inertia parameter or the gravitational parameter. As is easily recognized, sampling without error is impossible for all particle size ranges. It is, however, easy to find satisfactory sampling conditions, especially for fine particles.

3.2.3 Deposition of Particles in the Sampling Line

Some of the sampled particles will be deposited on the inside wall of the sampling line while they are transported from the sampling probe to a filter or a sample analyzer. The deposition efficiency is dependent on the particle size. Therefore, the data correction will be complicated. Some investigations have been carried out to prevent deposition by the use of a sheath airflow [22] or by feeding coarse particles into the sampling tube [23,24].

The main causes of the deposition are Brownian diffusion, gravitational sedimentation, and eddy diffusion. The details of these phenomena [25] are omitted here; only the deposition efficiencies are listed below.

Brownian Diffusion [26]

$$E = 1 - 0.819 \exp(-11.5\gamma) - 0.0976 \exp(-70.1\gamma)$$
$$- 0.0325 \exp(-179\gamma) + \cdots \quad \gamma \geq 0.01 \qquad (3.2.9a)$$

$$E = 5.50\gamma^{2/3} - 3.77\gamma - 0.814\gamma^{4/3} - \cdots \quad , \gamma < 0.01$$
$$(3.2.9b)$$

where

$$\gamma = \frac{4D_{BM}L}{\pi D^2 u} \qquad (3.2.10)$$

L (m) is the total length of the sampling tube and D (m) is the inside diameter of the tube. D_{BM} (m^2/s) is the particle diffusion coefficient given by the following equation:

$$D_{BM} = \frac{C_m kT}{3\pi\mu D_p} \qquad (3.2.11)$$

where k is Boltzmann's constant (1.3807×10^{-23} J/K), T (K) the temperature, and C_m (unitless) is Cunningham's slip correction factor. The slip correction factor C_m for a particle of diameter D_p (m) can be calculated by the following empirical equation:

$$C_m = 1 + \left[2.46 + 0.82 \exp\left(-0.44\frac{D_p}{\lambda}\right)\right]\frac{\lambda}{D_p} \qquad (3.2.12)$$

where λ (m) is the mean free path of the air molecule (6.46×10^{-8} m at 20°C, 1 atm).

62   Chapter 3

Gravitational Sedimentation [27,28]

$$E = 1 - \frac{2}{\pi} (\alpha\beta + \sin^{-1}\beta - 2\alpha^3\beta) \qquad (3.2.13)$$

where

$$\alpha = \left(\frac{3u_tL}{4uD}\right)^{1/3} \qquad (3.2.14a)$$

and

$$\beta = \sqrt{1 - \alpha^2} \qquad (3.2.14b)$$

The terminal settling velocity u_t with slip correction is given by the following equation:

$$u_t = \frac{C_m\rho_pD_p^2g}{18\mu} \qquad (3.2.15)$$

Eddy Diffusion [29,30]

$$E = 1 - \exp\left(-4\frac{\bar{V}^*L}{u^*D}\right) \qquad (3.2.16)$$

where $\bar{V}^*$ is the nondimensional deposition velocity averaged over the cross section of the sampling tube and u^* is the friction velocity. The deposition velocity $\bar{V}^*$ is a complicated function of the particle diameter, fluid friction factor, and so on [29]. The details are omitted here.

REFERENCES

1. T. Allen, Particle Size Measurement, 3rd ed., Chapman & Hall Ltd., London, Chap. 1 (1981).
2. K. Iinoya (ed.), Funtai Kogaku Binran (Powder Technology Handbook), Nikkan Kogyo, Tokyo, Chap. IV.4 (1986).

3. Japanese Industrial Standard M-8100, General Rules for Methods of Sampling of Bulk Materials (1973).

4. H. Masuda, H. Kurahashi, M. Hirota, and K. Iinoya, Kagaku Kogaku Ronbunshu 2, 286 (1976).

5. S. Miwa and J. Hidaka, Funtai Kogaku Jikken Manual (Japanese), Nikkan Kogyo, Tokyo, Chap. 1 (1984).

6. C. C. K. Wang, Chem. Eng. Prog. 80(9), 53 (1984).

7. F. Jordison, Chem. Eng. 85(24), 103 (1978).

8. P. G. Hoel, Elementary Statistics, John Wiley & Sons, Inc., New York (1966).

9. H. Masuda and K. Iinoya, J. Chem. Eng. Jpn. 4, 60 (1971).

10. H. Masuda, H. Yoshida, and K. Iinoya, J. Soc. Powder Technol. Jpn. 18, 177 (1981).

11. H. Yoshida, T. Ohsugi, H. Masuda, S. Yuu, and K. Iinoya, Kagaku Kogaku Ronbunshu 2, 336 (1976).

12. C. N. Davies, Br. J. Appl. Phys., Ser. 2, 1, 921 (1968).

13. P. A. Toynbee and W. J. S. Parkers, Int. J. Air Water Pollut. 6, 113 (1962).

14. I. Tamori, N. Kogure, and K. Imagami, J. Soc. Powder Technol. Jpn. 18, 165 (1981).

15 J. Bosch, Staub Reinhalt. Luft (English) 32, 8 (1972).

16. M. D. Durham and D. A. Lundgren, J. Aerosol Sci. 11, 179 (1980).

17. D. A. Lundgren and S. Calvert, Am. Ind. Hyg. Assoc. J. 28, 208 (1967).

18. J. H. Vincent, D. C. Stevens, D. Mark, M. Marshall, and T. A. Smith, J. Aerosol Sci. 17, 211 (1986).

19. Japanese Industrial Standard, Z 8808, Methods of Measuring Dust Content in Flue Gas (1977).

20. H. Masuda, H. Yoshida, and K. Iinoya, J. Chem. Eng., Jpn. 13, 467 (1980).

21. H. Yoshida, M. Uragami, H. Masuda, and K. Iinoya, Kagaku Kogaku Ronbunshu 4, 123 (1978).

22. M. B. Ranade, D. K. Werle, and D. T. Wassan, J. Colloid Interface Sci. 56, 43 (1976).

23. H. Masuda, S. Matsusaka, and Y. Sumiura, J. Soc. Powder
     Technol. Jpn. 20, 405 (1983).

24. N. Kogure, H. Yoshiyama, I. Tamori, T. Saito, and K. Wata-
     nabe, J. Soc. Powder Technol. Jpn. 23, 68 (1986).

25. S. K. Friedlander, Smoke, Dust and Haze, John Wiley &
     Sons, Inc., New York (1977).

26. P. Gormley and M. Kennedy, Proc. R. Irish Acad. A52,
     163 (1949).

27. N. A. Fuchs, The Mechanics of Aerosols, Pergamon Press,
     Ltd., Oxford, p. 112 (1964).

28. J. W. Thomas, Air Pollut. Control Assoc. J. 8, 32 (1958).

29. N. Yoshioka, C. Kanaoka, and H. Emi, Kagaku Kogaku 36,
     1010 (1972); 36, 1349 (1972).

30. S. K. Friedlander and H. F. Johnstone, Ind. Eng. Chem.
     49, 1151 (1957).

4

Powder Flow Rate

Methods for measurement of powder flow rate are discussed here
by classifying them into six groups: weighing method, flowmeters
based on mechanics, differential pressure method, electrical method, statistical method, and others. In Section 4.1 we outline
the principles of measurement, and in Section 4.8 we describe
several detectors of powder flow.

4.1 INTRODUCTION

There are two types of powder flow. One is bulk solids flow,
and the other is suspension flow as in the pneumatic transport
of pulverized coal. The principles of powder flowmeters for bulk
solids and suspension flow are listed in Tables 4.1 and 4.2, respectively. Weighing, listed in Table 4.1, belongs in the category of direct measurement, but the others are inferential measurements.

Table 4.1   Powder Flowmeter for Bulk Solids Flow

Principle	Note
Weighing	Belt scale, hopper scale
Impulsive force	Impact flowmeter
Coriolis force	Massometer[a] etc.
Back pressure of air jet	Gas-purge flowmeter
Volume displacement	Turbine flowmeter

[a]Trade name.

The powder flow rate W (kg/s) in suspension flow is the
product cvA (kg/s) of particle concentration c (kg/m^3), particle
velocity v (m/s), and the cross-sectional area A (m^2).   Therefore, the measurement of powder flow rate is replaced by a combined measurement of particle concentration and the particle velocity.   Although the measurement of the particle concentration is
discussed in Chapter 5, some of the methods for measurement are
also described in this chapter.

Table 4.2   Powder Flowmeter for Suspension Flow

Principle	Note
Coriolis force	Micromotion flowmeter[a] (slurry)
Drag force	Target flowmeter (slurry)
Pressure drop	Venturi flowmeter, venturi-orifice flowmeter
Electrostatic charge	Particle electrification by impact
Electromagnetic induction	Electromagnetic flowmeter (slurry)
Microwave	Microwave flowmeter (microwave resonance)
Statistical phenomena	Correlation flowmeter (concentration × velocity)
Heat absorption	Metallic particles
Volume displacement	

[a]Trade name.

Table 4.3 Particle Velocity Meter

Principle	Method
Statistical phenomena	Correlation technique (electrostatic current, electric capacitance, light extinction)
Spatial filtering	Microwave (standing wave) Laser velocimeter (lattice window)
Piezoelectric phenomena	Piezocrystal (momentum detector)
Doppler effect	Laser Doppler velocimeter
Time of flight	Double-beam laser

Table 4.3 shows the principles of the particle velocity measurement. The mean velocity of powder flow over the cross-
sectional area should be measured to obtain the powder flow rate
through the combined concentration-velocity method. However,
the laser Doppler velocimeter measures the particle velocity at a
point in the flow. Therefore, the velocimeter is not suitable for
the purpose of measuring the powder flow rate. The piezoelectric method is in the same category. These methods are used to
measure local powder flow rate. Although some particle velocity
meters have such problems in applications, they are also included
in this chapter.

4.2 WEIGHING METHOD

The weighing method [1] is the most reliable and accurate for
measurement of the powder flow rate in bulk solids flow. The
hopper scale (weigh hopper) and the belt scale (belt weigher or
conveyor scale) are the most famous flowmeters of this type.
Figure 4.1 shows the hopper scale. The scale receives powder
in the hopper, weighs it, and then discharges it. The frequency
of receiving and discharging is 100 to 400 cycles per hour, which
can be changed according to the process requirement. The weight
of powder received in one cycle is usually kept constant for the
sake of system simplicity. The hopper scale has the following
problems:

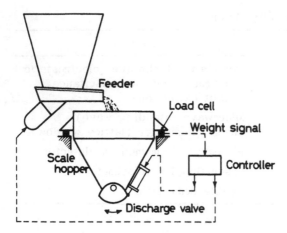

Figure 4.1   Hopper scale.

1. The powder flow rate is determined not by the preset weight,
    but by the actual weight received in the hopper.
2. The powder flow discharged from the hopper scale is inter-
    mittent. Therefore, the hopper scale is not a suitable sensor
    for the continuous process control.

   The powder flow from a feeder to the scale hopper should be
stopped as soon as the preset weight is reached. The powder
holdup between the feeder and the hopper is, however, an additional weight to the scale hopper, even if the powder flow is stopped instantaneously. There are two ways to avoid this error.
One is to reduce the feeding rate W (kg/s) when the weight is
reached to certain amount (say 90%) of the preset weight. Then
the mass of the holdup powder dw (kg) is reduced according to
the following simple relation:

$$dw = \frac{W}{v} \, d\ell \qquad\qquad\qquad (4.2.1)$$

where v is the mean velocity of powder and $d\ell$ is the distance between the feeder and the scale hopper. This "dribble" flow en-
sures a consistent cutoff, and a combination of high speed in the
early stage of feeding and low speed in the final stage allows a
short cycle time with high accuracy.

Fluctuations in feed rate also affect the accuracy of the hopper scale and may cause overshooting or premature cutoff for
the preset weight because of delay time $d\ell/v$ in the dynamic response [2]. The feeder should therefore maintain constant flow
rates with the greatest accuracy.

The other way to reduce the error is to weigh the scale hopper after the feeding flow is stopped completely. This is called
the net-weight mode. In this mode the hopper scale needs an
additional calculation unit and an additional idle period.

The second problem, intermittent flow discharged from the
hopper scale, is overcome partially by several modifications. The
first is to use two hoppers, as shown in Fig. 4.2. The timing
of feeding and discharge is controlled by a microprocessor or
microcomputer. Then the discharge flow becomes nearly constant.
The second modification is to follow the loss-in-weight mode, where
the powder flow rate is calculated by the following equation:

$$W(t) = -\frac{dG(t)}{dt} \qquad (4.2.2)$$

where $G(t)$ is the weight* of the hopper at time t.

If the powder is added before the hopper becomes empty, the
discharge flow is kept continuous. Further, the powder adhering
to the hopper wall does not cause a weighing error in the hopper
scale operated in this mode. Figure 4.3 shows a powder feeder
operated in the loss-in-weight mode. The weight of the feeder
decreases as

$$G(t) = G(0) - \int_0^t W(t)\, dt \qquad (4.2.3)$$

The weight $G(t)$ has to follow the straight line $G(t) = G(0) -$
kt in order to keep the flow rate $W(t)$ at a constant k (kg/s).
The feeder in Fig. 4.3 is controlled so as to follow the line. It

---

*Weight is the product of mass and the gravitational acceleration
g. The gravitational acceleration is omitted here for simplicity.
Therefore, $W(t)$ is the mass flow rate and $G(t)$ is, strictly speak-
ing, the mass of the hopper.

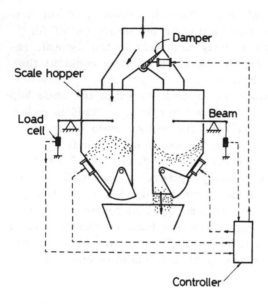

Figure 4.2 Hopper scale with two hoppers in parallel.

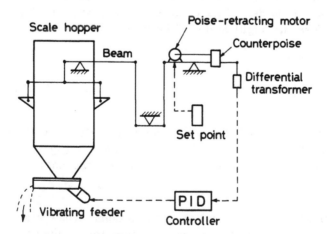

Figure 4.3 Hopper scale operated in the loss-in-weight mode.

should be clear that the small deviation of G(t) from the line
(reference line) causes a large fluctuation in the flow rate W(t).
Equation (4.2.3) shows that the hopper scale acts as an integral
element in the control system if the powder flow rate is the state
variable. In batch processes, however, the cumulative weight
G(0) − G(t) is sometimes more important than the instantaneous
flow rate W(t). The hopper scale works with high accuracy in
such a process.

For the continuous feeding of powder, the belt scale is more
suitable than the hopper scale. Figure 4.4 shows a belt scale
where the powder flow rate is obtained as a product of the weight
of powder on the belt of unit length and the belt speed. If the
flow rate is controlled so as to feed constant powder mass, the
belt scale is called a constant feed weigher. A mechanical unit
has been used to weigh the powder, and load cells have also been
utilized in recent years. The load cell, which is described in
Section 9.3, transforms the weight into an electric signal. A
roller pulley linked to a load cell supports the weight of the belt
and the powder. The weight acting on the pulley is, however,
changed by sag in the belt [3]. If the roller pulley (weigh idler)
is inserted between two pulleys as shown in Fig. 4.5, the load
suffered by the load cell should be half of the weight between

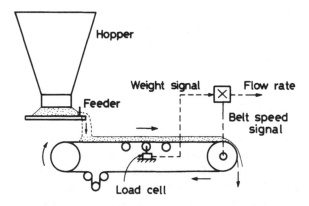

Figure 4.4   Belt scale.

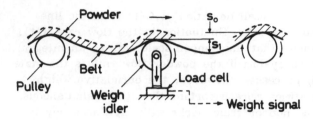

Figure 4.5 Weigh idler of a belt scale (the sags s_0 and s_1 are exaggerated).

the two pulleys. The error E caused by the belt sag is given by the following equations:

$$E = \frac{\sqrt{R}}{\sqrt{R} + \sqrt{R + 1}} - \frac{1}{2} \qquad \text{idler below pulley} \qquad (4.2.4a)$$

$$E = \frac{\sqrt{R}}{\sqrt{R} + \sqrt{R - 1}} - \frac{1}{2} \qquad \text{idler above pulley} \qquad (4.2.4b)$$

where R is defined by

$$R = \frac{\text{sag of the lowest part of the belt below the weigh idler}}{\text{sag of the weigh idler below or above the respective pulley}} = \frac{s_1}{s_0} \qquad (4.2.5)$$

If R becomes larger, the error E becomes smaller. If the weigh idler and the two pulleys are set at the same level ($s_0 = 0$), the ratio R is infinity and the error disappears.

The discussion above shows that the accuracy of the belt scale is more sensitive than the hopper scale to mechanical construction. Therefore, the belt scale may need frequent calibration. It is possible to weigh the entire belt scale to avoid the above-mentioned error. However, the method causes a dynamic delay in the response.

4.3 FLOWMETERS BASED ON MECHANICS

Powder flow causes various types of force. If the force is detected by a suitable method, the powder flow will be inferred.
Impulsive force and the Coriolis force are utilized in measurement
of the powder flow rate. The drag force on a target inserted in
the flow is also utilized in the measurement of slurry flow, because the drag coefficient is virtually constant in the range of
the flow Reynolds number between 10^4 and 10^7. These methods
are all inferential and need calibration except for the Coriolis
force flowmeter.

Figure 4.6 shows the principle of a flowmeter based on impulsive force, called an impact flowmeter. Powder falls by the force
of gravity and strikes the inclined plate. The force on the plate
can be calculated by the use of impulse-momentum relations. That
is, the normal component F_n is given by

$$F_n = W(v_{0n} + v_{1n})$$ (4.3.1a)

and the parallel component F_p is given by

$$F_p = W(v_{0p} - v_{1p})$$ (4.3.1b)

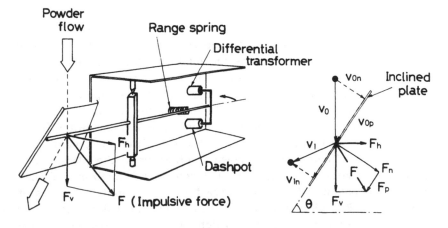

Figure 4.6   Impact flowmeter.

where v is the powder impact velocity and the subscripts 0 and 1 refer to conditions before and after impact.

The impact velocity v_0 is given by the following equation:

$$v_0 = k\sqrt{2gh} \qquad\qquad (4.3.2)$$

where h is the free-falling height. The coefficient k, which represents air drag, is less than unity.

Equation (4.3.2) represents the free-falling powder velocity very well, even if the powder is fine, and the coefficient k is nearly equal to unity. The velocities v_{1n} and v_{1p} are represented by the following two equations, respectively:

$$v_{1n} = ev_{0n} \qquad\qquad (4.3.3a)$$

$$v_{1p} = \mu v_{0p} \qquad\qquad (4.3.3b)$$

where e is the coefficient of restitution and μ is the coefficient of friction. From Eqs. (4.3.1) to (4.3.3), the force on the inclined plate is given by the following equations:

$$F_n = k(1 + e)\cos\theta\,\sqrt{2gh}\,\,W \qquad\qquad (4.3.4a)$$

$$F_p = k(1 - \mu)\sin\theta\,\sqrt{2gh}\,\,W \qquad\qquad (4.3.4b)$$

where θ is the inclination of the plate.

The force obtained will also be represented by the horizontal and vertical components. The impact flowmeter detected the vertical component in the early stage of development [4]. The vertical force detected is, however, the sum of the impulsive force and the weight of the powder adhering to the plate. Therefore, the flowmeter indicates a certain value even if the powder flow rate is zero. On the other hand, the horizontal component of the force is not affected by the powder adhering to the plate. The impact flowmeter shown in Fig. 4.6 detects the horizontal force through a balance of the impulsive force and repulsive forces of a spring. The displacement resulting from the balance is detected by a differential transformer, which gives an electric signal pro-

portional to the displacement (see also Section 9.3.2). From Eq.
(4.3.4), the horizontal component of the impulsive force is given
by the following equation:

$$F_h = F_n \sin \theta - F_p \cos \theta$$

$$= \frac{1}{2}(e + \mu)k \sin 2\theta \sqrt{2gh}\ W \qquad (4.3.5)$$

The coefficient of restitution and the coefficient of friction
in Eq. (4.3.5) may depend on the powder properties, such as
particle size and particle shape. Therefore, calibration is necessary for the impact flowmeter. Abrasion of the inclined plate
may change the direction of impact force, resulting in measurement error. However, the flowmeter is very compact and the dynamic response very fast. Further, dust is easily kept out of
mechanical and electrical parts. Therefore, the flowmeter is suitable for continuous control of powder handling processes, and is
used widely throughout the world.

Similar flowmeters based on impulsive force are commercially
available. However, it is important to know whether the force
detected is vertical or horizontal, as mentioned above.

This principle is also applied to gas-solid suspension flow
[5]. If the flow strikes a target or a suspended dram as shown
in Fig. 4.7, the impulse-momentum relation gives the following
equation:

$$F = \rho A u \cdot u + W(v_0 + v_1)$$

$$= Qu + Wv(1 + e) \qquad v = v_0 \qquad (4.3.6)$$

where ρ is the gas density, u the gas velocity, and Q the gas
flow rate (kg/s). From Eq. (4.3.6), the powder flow rate W is
given by the following equation:

$$W = \frac{F - Qu}{v(1 + e)} \qquad (4.3.7)$$

The velocities u and v in Eq. (4.3.7) should be measured by a
suitable method. Also, the particle inertia must be large enough

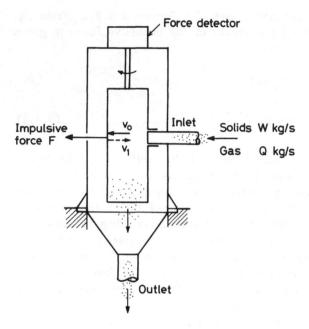

Figure 4.7   Principle of the impact flowmeter for suspension flow.

so that all the particles strike the dram.   The measurement of gas flow rate in gas-solid suspension flow is discussed in Section 4.4.

The next subject concerns the Coriolis force flowmeter.   The Coriolis force is a fictitious force found in a rotating frame [6]. Another fictitious force found in a rotating frame is the well-known centrifugal force.   If the powder flow is received on a rotating disk and thrown off from the periphery as shown in Fig. 4.8, the powder on the disk experiences a Coriolis force in the counterdirection of the rotation.   The force dF_c acting on the holdup powder dw is given by the following equation:

$$dF_c = 2\omega v \ dw \qquad\qquad\qquad (4.3.8)$$

where ω is the angular velocity of the disk and v is the radial velocity of powder on the disk.   The holdup Powder dw is given by (W/v) dr as in Eq. (4.2.1).   Therefore, Eq. (4.3.8) reads

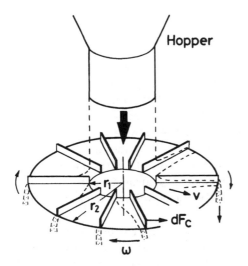

Figure 4.8 Principle of a Coriolis force flowmeter.

$$dF_c = 2\omega W \, dr \tag{4.3.9}$$

The force acts on the disk because the powder flow between r_1 and r_2 is restricted by the blades to the radial direction. The resulting torque on the rotating axis is given by the following equation:

$$T_1 = \int_{r_1}^{r_2} r \, dF_c = 2\omega W \int_{r_1}^{r_2} r \, dr$$

$$= \omega(r_2^2 - r_1^2)W \tag{4.3.10}$$

Further, the blades on the disk experience an impulsive force at $r = r_1$, which is caused by the change of the flow direction. As in Eq. (4.3.1), the impulsive force F_i is given by

$$F_i = Wr_1\omega \tag{4.3.11}$$

Then the total torque T is given by the following equation:

$$T = T_1 + F_i r_1$$

$$\quad = \omega r_2^2 W \qquad\qquad\qquad\qquad\qquad (4.3.12)$$

The Coriolis force flowmeter detects the torque T.

Equation (4.3.12) contains no calibration constant. This is the most important feature of the flowmeter based on the Coriolis force. Figure 4.9 shows examples of the flowmeter. However, these flowmeters are not as widely utilized in industry as is the impact flowmeter. There may be some trouble with the rotating parts and flow distortion or blockage may arise in actual applications. Therefore, further modifications will be necessary to overcome these difficulties.

The Coriolis force is also found in a field oscillating with angular frequency ω. As shown in Fig. 4.10, if a U-tube through which a fluid flows is oscillated around an axis connecting the inlet and outlet, the U-tube will be twisted by the Coriolis force. The oscillation will be carried out by use of a spring and an electromagnet. The Coriolis force caused by the inflow is given by $2\omega QL$, where Q is the fluid flow rate and L the length of the

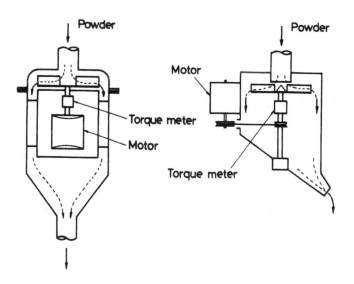

Figure 4.9 Coriolis force flowmeters.

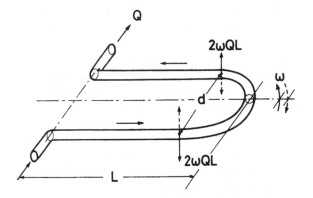

Figure 4.10   Principle of a Coriolis force flowmeter in vibration
mode.

U-tube.   This force acts on the U-tube in the direction counter
to the oscillation.   On the other hand, the Coriolis force by the
return flow is also given by $2\omega QL$.   This force acts on the U-tube
in the direction of the oscillation.   Therefore, the torque around
the axis of the U-tube is given by $2\omega QLd$, where d is the normal
distance between the inlet and outlet of the U-tube.   If the torsional rigidity of the U-tube is denoted by I_r and the torsional
angle by θ, the following equation is obtained:

$$2\omega QLd = I_r \theta \tag{4.3.13}$$

Therefore, if the torsional angle θ is measured, the mass flow
rate Q will be obtained.
    The oscillating U-tube is applicable to the slurry flow but not
to the powder flow or gas-solid suspension flow.   The powder
flow will block the U-tube.   The particles in the gas-solid suspension do not effectively transfer the Coriolis force to the U-tube
unless the flow is extremely dense.

4.4   DIFFERENTIAL PRESSURE METHOD

Mass flowmeters based on the differential pressure or pressure
drop are applied primarily to gas-solid suspension flow.   However,

the principle is also utilized in the measurement of powder flow, where the flowmeters described in Sections 4.2 and 4.3 are not applicable.

The differential pressure method has the advantage that the electric parts of the flowmeter can be isolated from the main process. The measurement system for gas pressure is, however, affected by the particles or by dust. This problem is discussed in Section 9.4.

When the particles are transported by a pneumatic conveyor, the pressure drop along the conveyor line is usually larger than that of pure airflow. The additional pressure drop depends on the mass flow ratio m, which is defined by

$$m = \frac{\text{powder flow rate}}{\text{gas flow rate}} = \frac{W}{Q} = \frac{W}{\rho u A} \qquad (4.4.1)$$

In the measurement of powder flow rate, it is necessary to know the relationship between the mass flow ratio and the pressure drop together with the gas flow rate in the gas-solid suspension flow. It will be shown here that a venturi or orifice meter is applicable for this purpose. A venturi diffuser is preferable because an orifice may be severely eroded and disturb the flow, and fine particles may deposit on the corner parts. Therefore, the following discussion is concerned with the diffuser [7].

Assuming the incompressibility of gas, the velocity u in a diffuser is given by the following equation:

$$u = \frac{u_0}{(1 + \alpha x)^2} \qquad (4.4.2)$$

where u_0 is the gas velocity at the diffuser inlet, x the distance from the inlet in the flow direction, and $\alpha = \tan \theta / r$ (r is the inner radius of the diffuser inlet pipe, and θ is half of the diverging angle of the diffuser).

The gas velocity decreases with distance x as shown by Eq. (4.4.2). However, the particle velocity will not decrease as rapidly because of the particle inertia. The particle velocity v_0 at the inlet is smaller than the gas velocity u_0 because the particles have suffered wall friction. Therefore, the velocity increases from v_0 to the maximum value in the diffuser, where the friction disappears. Then the velocity gradually decreases from the maximum to the gas velocity. It is assumed in the following analysis that the particle is small. The measurement error for larger particles will be smaller than that estimated in the following analysis. The equation of motion for the small particle is represented as follows:

$$m_p \frac{dv}{dt} = 3\pi\mu D_p (u - v) \qquad (4.4.3)$$

where m_p is the mass of the particle $[= (\pi/6)D_p^3 \rho_p]$, D_p the particle diameter, ρ_p the particle density, and μ the viscosity of the gas. Equation (4.4.3) is rewritten with the relation $v = dx/dt$ as follows:

$$\tau \frac{dv}{dx} = \frac{u}{v} - 1 \qquad (4.4.4)$$

The parameter τ is called the particle relaxation time, which is defined by

$$\tau = \frac{\rho_p D_p^2}{18\mu} \qquad (4.4.5)$$

The approximate solution of Eq. (4.4.4) is given by the following equation:

$$v = v_0 + \frac{x(1 - \phi_0 - \alpha\phi_0 x)}{\tau\phi_0 (1 + \alpha x)} \qquad (4.4.6)$$

where, $\phi_0 = v_0/u_0$. Therefore, x_0, where the particle velocity v is equal to v_0, is given by

$$x_0 = \frac{1 - \phi_0}{\alpha\phi_0} \qquad (4.4.7)$$

From the force balance at any axial position x, the following relation will be obtained:

$$dp_s \ A = dw \ \frac{dv}{dt} \tag{4.4.8}$$

where dp_s is the additional pressure drop by particles, A the cross-sectional area at position x, and dw the mass of the holdup powder between x and x + dx.

The holdup dw is given by (W/v) dx, as in Eq. (4.2.1). Therefore, Eq. (4.4.8) is rewritten with mass flow ratio m:

$$dp_s \ A = \frac{W}{v} \ dx \ \frac{dv}{dt} = m\rho A \ \frac{u}{v} \ v \ dv \tag{4.4.9}$$

and Eq. (4.4.9) is simplified to

$$dp_s = m\rho u \ dv \tag{4.4.10}$$

On the other hand, the pressure difference by the gas flow is given by the following equation:

$$dp_a = \rho u \ du \tag{4.4.11}$$

If the additional pressure difference Δp_s is equal to zero, the gas flow rate Q is determined by

$$Q = \rho u A = K\rho \ \sqrt{\frac{\Delta p_a}{\rho}} \tag{4.4.12}$$

where K is the calibration constant. For the case of $\Delta p_s \neq 0$, the relative error caused by the additional pressure difference is given by

$$\frac{Q_m - Q}{Q} = \sqrt{1 + \frac{\Delta p_s}{\Delta p_a}} - 1 \cong \frac{1}{2} \frac{\Delta p_s}{\Delta p_a} \tag{4.4.13}$$

where Q_m is the measured gas flow rate. From Eqs. (4.4.10) and (4.4.11), the following equation will be obtained:

$$\frac{\Delta p_s}{\Delta p_a} = m \ \frac{\int u \ dv}{\int u \ du} \tag{4.4.14}$$

From Eq. (4.4.10), the additional pressure drop Δp_s at position x_0 where the particle velocity is equal to v_0 is given by

$$\Delta p_s = \int_{v_0}^{v_0} m\rho u \, dv = 0 \qquad (4.4.15)$$

Therefore, the effect of particles on the pressure difference of gas-solid suspension flow disappears at position $x = x_0$. The pressure difference ratio $\Delta p_s/\Delta p_a$ near the position x_0 is calculated by the use of Eqs. (4.4.14), (4.4.2), and (4.4.7) as follows:

$$\frac{\Delta p_s}{\Delta p_a} = \frac{m}{\tau u_0} \frac{\phi_0^2}{1 + \phi_0 + \phi_0^2 + \phi_0^3} (x - x_0) \qquad (4.4.16)$$

If τu_0 and $x - x_0$ are divided by D (= $2r$), Eq. (4.4.16) becomes a dimensionless form. The dimensionless parameter $\tau u_0/D$ is called the inertia parameter (= ψ), that is,

$$\psi = \frac{\tau u_0}{D} = \frac{\rho_p D_p^2 u_0}{18 \mu D} \qquad (4.4.17)$$

Equations (4.4.13) and (4.4.16) show that the relative error is directly proportional to the mass flow ratio m and inversely proportional to the inertia parameter. The error is smaller for larger particles and for a higher velocity or smaller diameter of diffuser inlet.

The more precise value of position x_0 will be obtained by numerical integration of the equation of motion. Also, the pressure difference ratio will be obtained for any position x through numerical integration of Eq. (4.4.14). Then the measured gas flow rate will be corrected by use of Eq. (4.4.13).

The analysis for the orifice meter is similar to the above, and the error becomes smaller as the inertia parameter becomes larger.

Figure 4.11 shows a venturi-type flowmeter. The pressure drop Δp is represented by [see also Eq. (4.4.10) and (4.4.11)]

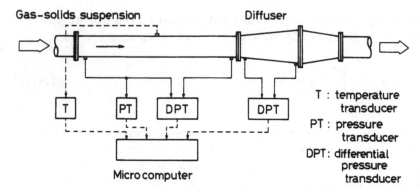

Figure 4.11 Venturi flowmeter.

$$\Delta p = m\rho \int u\ dv + \rho \int u\ du + \frac{\lambda_s^m}{2D} \int uv\ dx + 4f\ \frac{\rho u^2}{2}\ \frac{\Delta x}{D}$$

$$(4.4.18)$$

where λ_s and f are the friction factors for particles and air, respectively. For fully developed pipe flow, $du = dv = 0$. Further,
the product uv in the third term can be replaced by ϕu^2, because the velocity ratio v/u is assumed to be constant ϕ.

On the other hand, the last term in Eq. (4.4.18) is represented by the pressure difference at the diffuser as follows:

$$4f\ \frac{\rho u^2}{2}\ \frac{\Delta x}{D} = k\ \Delta p_a$$

$$(4.4.19)$$

Further, $\lambda_s \phi / 4f$ is assumed to be a constant k_1. Therefore, the
mass flow ratio m is given by the following equation:

$$m = \frac{\Delta p - k\ \Delta p_a}{kk_1\ \Delta p_a} = K' \frac{\Delta p - k\ \Delta p_a}{\Delta p_a}$$

$$(4.4.20)$$

where K' and k are the calibration constants. From Eqs. (4.4.12)
and (4.4.20), the powder flow rate W is calculated by

$$W = K'' \sqrt{\rho} \; \frac{\Delta p - k \Delta p_a}{\sqrt{\Delta p_a}} \tag{4.4.21}$$

The calibration constant k is adjustable to unity if the distance between pressure taps in the straight pipe section is determined so that the pressure drop for pure airflow is equal to the pressure recovery of the diffuser.

The pressure drop by a venturi nozzle is utilized in the venturi-orifice flowmeter [8]. The analysis is similar to that above. Equation (4.4.18) is also valid in this case. The particle velocity is, however, accelerated by airflow and the assumption of the constant velocities u and v, and the constant velocity ratio v/u may not be valid. Therefore, the calibration constant K'' will depend on the derivative dv/du.

A venturi with a long throat is also used to increase the pressure drop and the meter sensitivity [9]. Although Eq. (4.4.18) is also valid in this case, further analysis will be necessary on the particle velocity change. Another version of the flowmeter is composed of a straight pipe and a core inserted so as to make an annular flow between the pipe and the core. The particles will be accelerated by the high-speed annular flow, resulting in the pressure drop. The flowmeter is affected less by upstream disturbances and does not require such a long, straight inlet pipe as in the case of the venturi flowmeter.

The flowmeters discussed above will be used in gas-solid suspension flow with mass flow ratio below 10. For higher mass flow ratio, the relation between the pressure drop ratio ($\Delta p / \Delta p_a$) and the mass flow ratio m becomes nonlinear. However, the following equation will be used in a restricted measurement range:

$$m = m_0 + \left(\frac{\partial m}{\partial a} \right)_0 da + \left(\frac{\partial m}{\partial u} \right)_0 du \tag{4.4.22}$$

where a is the pressure drop ratio. The flowmeter shown in Fig. 4.11 utilizes a microcomputer in order to calculate the mass flow ratio and powder flow rate following Eqs. (4.4.20) to (4.4.22), and to correct the variation of air density with the static pressure and the temperature.

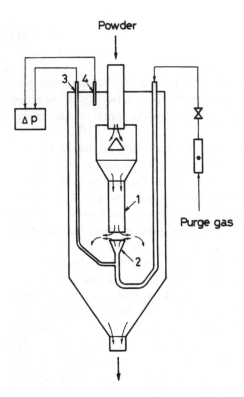

Figure 4.12   Gas-purge flowmeter (back pressure of an air jet
is detected).

The differential pressure method is also utilized in the measurement of powder bulk flow.   Figure 4.12 shows a gas purge
flowmeter [10].   The meter will be used even if the nuclear radiation is high enough to damage the electrical insulations, because the electric parts are easily isolated from the process.
The powder flows down in the rectifying pipe (1) by gravity,
and is decelerated by a counter gas flow purged from the nozzle
(2).   The purge gas flow rate is held constant by use of a flow
regulator.   Then the pressure difference between tap 3 and tap
4 increases with the powder flow rate.   The force balance equation, similar to Eq. (4.4.8), is also valid in this case.

$$A \, dp = -W \, dv \qquad\qquad (4.4.23)$$

The particle velocity v in the equation will be obtained by numerical integration of the equation of motion. Experimental data show the linear relationship between the powder flow rate and the pressure difference, as long as the powder flow rate is not so high.

4.5 ELECTRICAL METHOD

The electrical phenomena utilized in the measurement of powder flow rate are electrostatic charging, piezoelectric effect, the electric capacitance increase by particles, and microwave resonance. Induction charging is also utilized for electroconductive particles such as metallic powder or liquid droplets. Further, the electromagnetic induction is applied to the slurry measurement.

4.5.1 Electrostatic Charging

Static electrification of particles takes place in various types of powder handling processes. The cause of electrification is contact or impact between particles and the walls of the equipment. If a particle strikes a metallic body, the particle and the metal acquire an opposite charge of equal amount, respectively. The charge on the metal is measured as an electric current by grounding the metal as shown in Fig. 4.13.

The driving force of the charge transfer is the difference in electron energy between the two bodies: the particle and the metal. The representative electron energy of a material is given by a Fermi level, which depends on the type of material. The amount of charge transferred is proportional to the difference in the Fermi levels. The energy required to pull up an electron from the Fermi level to the vacuum level (outside the body) is called a work function or chemical potential. Then the charge transferred is proportional to the following quantity:

$$V_c = \frac{\phi_s - \phi_m}{e} \tag{4.5.1}$$

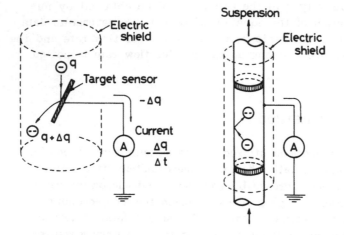

Figure 4.13 Electrostatic charging by impact on metallic sensor.

where ϕ_s and ϕ_m (J) are the work functions for the particle and
the metal, respectively, and e is the elementary charge (= 1.6 ×
10^{-19} C). V_c (V) is called the contact potential difference.

The charge transferred Δq (C) is represented by the following equation:

$$\Delta q = \frac{S V_c \varepsilon_0}{z_0 \rho_d \varepsilon_s} \Delta t \qquad (4.5.2)$$

where S is the contact area, ε_0 the dielectric constant of air
(= 8.85 × 10^{-12} F/m), ε_s the dielectric constant of the particle,
z_0 the gap between contact bodies ($\cong$ 4 Å), ρ_d ($\Omega \cdot$ m) the specific resistance of the particle, and Δt (s) the contact time.

In the derivation of Eq. (4.5.2), it is assumed that the contact time is very short compared with the time constant of charge
transfer $\rho_d \varepsilon_s$. A very high concentration of the surface states*
is also assumed.

---

*The surface of a particle is, in some cases, different from the
body. The electron-energy level is also different for the surface.
The number of active sites on the surface per unit area and unit
energy level is called the surface state.

From the conservation of electric charge, the current I(A) generated on the metal is given by

$$I = - \frac{\Delta q}{m_p} W = - \frac{S V_c \, \Delta t \, \epsilon_0}{m_p z_0 \rho d^\epsilon s} W \qquad (4.5.3)$$

The contact area S is proportional to D_p^2 and the mass of the particle m_p is proportional to D_p^3. Therefore, the current I is inversely proportional to the particle size. The contact area also depends on the impact velocity [11.12], and so does the current. Equation (4.5.3) can be applied to the measurement of powder flow rate W both in gas-solid suspension flow and bulk solids flow. However, the flow rate must be rather small, because the relation between the current and the flow rate becomes nonlinear for higher flow rate [13]. Also, the impact becomes ineffective if the metal is covered by particles already in contact.

The current generated becomes large if the metal is covered by plastics or ceramics. Then the sensitivity becomes high enough to measure the small flow rate of particles, such as the dust flowing in a smoke duct.

The problems with measurement by the use of electrostatic charging are as follows:

1. The current depends on the materials; the surface condition will gradually be changed by the impact.
2. The current depends on the particle size.
3. The current depends on the impact velocity.
4. The current depends on the initial charge of particles.

4.5.2 Piezoelectric Effect

The method discussed above will also be applied to measurement of the local particle flow rate. The piezoelectric method described here is used only to detect local flow [14—16]. The principle is quite different from that of electrostatic charging. The sensor is made of a piezoelectric crystal or ceramic, which are very cheap and widely used as igniters. If the dielectric crystal is deformed by an impulsive force, ions in the crystal will be displaced as shown in Fig. 4.14. The displacement of ions is called polarization. Polarization induces an electric pulse between the upper

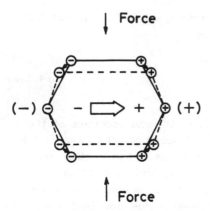

Figure 4.14 Polarization of piezoelectric crystal.

and lower surfaces of the crystal. The pulse will be detected if
the surfaces are coated by metal and connected to a pulse analyzer. Figure 4.15 shows a piezoelectric sensor. The impulsive
force is given by Eq. (4.3.1a), and the following equation will be
obtained:

$$\int_0^\infty H(t) \, dt = km_p(1 + e)v_{0n} \qquad\qquad (4.5.4)$$

where $H(t)$ is the electric pulse, v_{0n} the impact velocity normal
to the sensor, and k the proportional constant (1 to 2 V/N [15]).

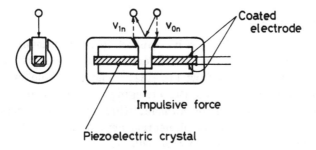

Figure 4.15 Piezoelectric sensor for the measurement of particle
momentum.

The integral $\int H(t)\, dt$ may be proportional to the initial pulse height $H(0)$. The electronic circuitry used to analyze the pulses is given in the literature [16].

The basic equation (4.5.4) is, however, not utilized in measurement of the local mass flux. The flux J $(kg/m^2 \cdot s)$ is given by

$$J = \frac{m_p n}{\eta A} \qquad\qquad (4.5.5)$$

where n is the number frequency of impact (s^{-1}), A the sensor area (m^2), and η the impact efficiency (unitless). The number frequency of impact n is determined by the number of electric pulses per second detected by the piezoelectric sensor.

The impact efficiency η for suspension flow is a function of the inertia parameter defined by Eq. (4.4.17) with appropriate length D and velocity u_0 [17].

Further, the local mass concentration c (kg/m^3) is given by the following equation:

$$c = \frac{J}{v_{0n}} \qquad\qquad (4.5.6)$$

The equations above show that the particle mass should be known to measure the local mass flux or concentration by use of the piezoelectric sensor.

A strain gauge described in Section 9.3.1 can also be utilized as a sensor of an impact probe instead of a piezoelectric crystal [18].

4.5.3 Electric Capacitance Method

The following method is, in principle, applied to the measurement of volumetric concentration of particles. However, the powder flow rate will also be obtained as far as the powder velocity is known.

If a powder falls down by gravity between metallic parallel plates as shown in Fig. 4.16, the capacitance C is given by the following equation:

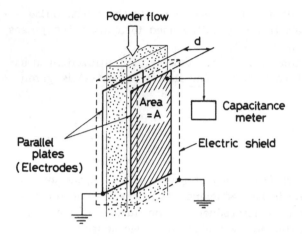

Figure 4.16   Electric capacitance method for the measurement of
volumetric concentration of particles.

$$C = \frac{\varepsilon A}{d}$$

(4.5.7)

where A is the area of the plate, d the distance between plates,
and ε the dielectric constant of gas-solid suspension between the
plates.   The dielectric constant of the suspension is usually larger
than the dielectric constant of air ε_0 (= 8.85 × 10^{-12} F/m).   If
there are no particles between the plates, ε is equal to ε_0.   An
increase in dielectric constant $\Delta\varepsilon$ gives the increased capacitance
ΔC.

The following equation gives the dielectric constant for a
packed bed of spherical particles (Rayleigh's equation [19]):

$$\frac{\varepsilon}{\varepsilon_0} = 1 + \frac{3c_v}{\dfrac{\varepsilon_s + 2\varepsilon_0}{\varepsilon_s - \varepsilon_0}\, c_v - 1.65 \dfrac{\varepsilon_s - \varepsilon_0}{\varepsilon_s + \frac{4}{3}\varepsilon_0}\, c_v^{10/3}}$$

(4.5.8)

where c_v is the volumetric concentration:*

---

*The concentration c is the holdup concentration (= c_h).   See
Section 5.3 for a definition of particle concentration.

$$c_v = \frac{c}{\rho_p} \qquad (4.5.9)$$

If the volumetric concentration is much smaller than unity, the following equation will be obtained from Eq. (4.5.8):

$$\frac{\Delta\varepsilon}{\varepsilon_0} = \frac{\varepsilon}{\varepsilon_0} - 1 = \frac{3(\varepsilon_s - \varepsilon_0)}{\varepsilon_s + 2\varepsilon_0} c_v \qquad (4.5.10)$$

Further, if ε_s is much larger than ε_0, Eq. (4.5.10) is simplified to

$$\frac{\Delta\varepsilon}{\varepsilon_0} = 3c_v \qquad (4.5.11)$$

Equations (4.5.10) and (4.5.11) show that an increase in the capacitance ΔC is proportional to the volumetric concentration. If the dielectric constant of particles is much larger than that of the air, the increase in capacitance does not depend on the materials, as shown by Eq. (4.5.11). Therefore, the volumetric concentration of metallic particles ($\varepsilon_s \cong \infty$) will be measured independent of material type. The shape of particles, however, may affect the increase of the capacitance.

Equation (4.5.10) will be applied to the measurement of powder flow rate even for nonmetallic particles. In this case, the calibration constant for the measurement will depend on the dielectric constant of the particles. Therefore, the constant will vary with the type of powder. The characteristic is not preferable for the measurement. This fact, however, suggests further application of the method in the measurement of a mixing ratio composed of two types of powder. If the powder flow rate is measured by other flowmeters, such as the impact flowmeter discussed in Section 4.3, the mixing ratio will be inferred by use of Eq. (4.5.10) with linear interpolation [20].

The volumetric concentration in gas-solid suspension flow with mass flow ratio m below 10 is usually too low to detect an increase in capacitance, because the density of particles is much greater than that of the air, and therefore the volumetric concentration is very low [see also Eq. (4.4.1)]. For a coal-air mixture, the following equation was obtained [21]:

$$\log \frac{\varepsilon}{\varepsilon_0} = c_v \log \frac{\varepsilon_s}{\varepsilon_0} \qquad (4.5.12)$$

Equation (4.5.12) shows a nonlinear relationship between the volumetric concentration and the dielectric constant. The plate used in the experiment was about 3 mm × 180 mm, and the distance d was about 3 mm. Therefore, the duct for the flow was a small square. The experimental fact suggests that the capacitance depends on the geometry of the electrodes even if they are set in parallel.

The electronic circuit for the measurement of the capacitance is given in Section 8.4.1. Also, a C-meter or Q-meter is utilized in the measurement [21].

The dielectric constant of particles ε_s is determined by use of standard liquids such as benzene or acetone. The dielectric constant of the liquid is changed by adding another liquid. If the capacitance is not changed by immersion of particles into the liquid, the dielectric constant of particles is equal to that of the liquid. Therefore, the dielectric constant of particles will be obtained. The specific dielectric constant (the dielectric constant divided by that of air, $\varepsilon_s / \varepsilon_0$) of particles thus measured is usually between 2 and 10, except for the metallic particles. On the other hand, the specific dielectric constant of water is about 80 and therefore much larger than that of particles. Then the water content of powder will affect measurement of the volumetric concentration (see also Section 8.4.1).

4.5.4 Electromagnetic Induction

The electromagnetic induction is known as Faraday's law. If an electric coil is moved in a magnetic field, a so-called electromotive force will be induced in the coil. The electromotive force is proportional to a decrease in the magnetic flux through the coil. Figure 4.17 shows an electromagnetic flowmeter. The pipe is made from nonmagnetic material such as stainless steel, through which the magnetic field can pass. The inside wall of the pipe is covered by a nonconductive plastic or ceramic. A pair of electrodes is set on the wall perpendicularly to the magnetic field. The charge carriers (electron or ion) in the conductive fluid experience an electromagnetic force in the direction of the electrodes and form a charge distribution. Therefore, an electric

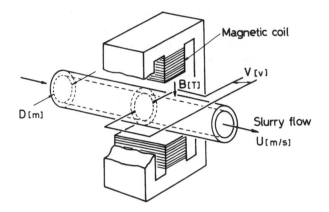

Figure 4.17   Electromagnetic flowmeter.

field is formed in the fluid by the charge distribution. The
charge carriers are also affected by the electric field, resulting
in an equilibrium distribution. The electric field thus formed
causes a potential difference between the electrodes. The potential difference V (V) is given by the following equation:

$$V = kBDu \qquad\qquad (4.5.13)$$

where k is the proportional constant (unitless), B the magnetic
flux density [T (= $kg \cdot s^{-2} A^{-1}$)], D the inside diameter of the
pipe (m), and u the mean flow velocity (m/s). The volumetric
flow rate Q_V(m^3/s) is given by

$$Q_v = \frac{\pi D}{4kB} V \qquad\qquad (4.5.14)$$

The electromagnetic flowmeter is truly obstructionless and
widely utilized in measurement of the flow rate of fluids with
a specific resistance smaller than $10^4 \, \Omega \cdot m$. If the resistance is
too high, the potential difference is hardly detected. The air
has a very high resistance, and the flowmeter is not applicable.
It is, however, also utilized in the measurement of slurries where
the particles are uniformly suspended. The meter is calibrated
by the use of water for a low concentration, because the potential
difference does not depend on the temperature, density, and
viscosity of the fluid. The particle flow rate in the suspension

will be obtained as a product of the concentration and the volumetric flow rate of the suspension. If the concentration is higher than 10% the proportional constant may become larger because of the increase in magnetic flux density, and recalibration of the flowmeter will be necessary. Also, the particle deposition on the wall will affect the pickup of potential difference, resulting in a measurement error.

The magnetic field is given by an alternating current rather than a direct one, because the signal amplification is easy and the potential difference is not affected by electrolytic polarization of the fluid.

4.5.5 Microwave Resonance

The microwave is an electromagnetic wave with wavelength between 0.03 and 30 cm. The frequency of the wave is between 10^9 and 10^{12} Hz. The electromagnetic wave will be represented by Maxwell's equations, which are general equations including Faraday's induction law. If the equations are solved under some boundary conditions, primitive solutions will be obtained. The solutions for a circular tube resonator depend on the integers m and n. The wave propagation mode corresponding to each primitive solution is called the normal mode, and denoted by abbreviated symbols such as TM_{mn}, TE_{mn}, or HE_{mn}. The TM mode is the transverse magnetic mode, where there is no magnetic field component in the direction of wave propagation. On the other hand, the TE mode is the transverse electric mode, where there is no electric field component in the direction of wave propagation. If the wave has both magnetic and electric field components in the direction of propagation, it is called the hybrid mode. The HE mode is one of the hybrid modes, where the electromagnetic field is approximately equal to that of the TE mode.

A waveguide usually has a cutoff frequency, and only the higher-frequency wave can pass through the guide. The wave that has the lowest cutoff frequency is called the dominant mode.

Figure 4.18 shows a circular tube resonator lined with a dielectric material [22]. The wavelength of the microwave is determined so that the wave in the resonator propagates in the HE_{11} mode. The wavelength determined is comparable to 2b,

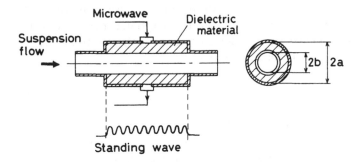

Figure 4.18 Microwave resonator.

where b is the inside radius of the resonator. The dominant
mode in this case is the HE_{11}, and the mode separation is easier
than the other modes. On the other hand, the ratio a/b of the
outside and inside diameters of the dielectric liner is determined
so that the electric field formed in the resonator becomes uniform.

If the gas-solid suspension flows through the resonator, the
resonant frequency changes from ω_0 to ω. The relative decrease
in the frequency is given by the following equation:

$$\frac{\Delta\omega}{\omega_0} = \frac{\omega}{\omega_0} - 1 = - \frac{\int_{\Delta V} \Delta\varepsilon_s |E|^2 \, dV}{4U}$$

$$\cong - \frac{\Delta\varepsilon_s |E|^2}{4U} \, \Delta V \qquad\qquad (4.5.15)$$

where E is the electric field component, ΔV the total volume of
the particles in the resonator, $\Delta\varepsilon$ the difference in the dielectric
constant $\varepsilon_s - \varepsilon_0$, and U the time-averaged electric energy:

$$U = \frac{1}{2} \int_V \varepsilon_0 |E|^2 \, dV \qquad\qquad (4.5.16)$$

Therefore, the volumetric concentration c_v (= $\Delta V/V$, V is the
volume of the resonator) will be obtained by the use of Eq.
(4.5.15).

On the other hand, there exists a standing wave in the resonator as shown in Fig. 4.18. If the particles pass through the wave with velocity v, an electric signal with the following frequency f_v (Hz) will be superposed:

$$f_v = \frac{v}{\lambda} \qquad\qquad (4.5.17)$$

where λ is the wavelength of the standing wave.

Equation (4.5.17) shows that the standing wave works as a lattice window. The method for measurement of velocity by use of a lattice window is called spatial filtering.

The particles have to be shaped into clouds in order to utilize the spatial filtering, because the uniform concentration distribution of particles gives no signal change even if the standing wave exists. Fortunately, the distribution of particles cannot be uniform because of the turbulence and the interference between the particles.

From Eqs. (4.5.15) and (4.5.17), the powder flow rate will be measured by the system shown in Fig. 4.19. The resonance

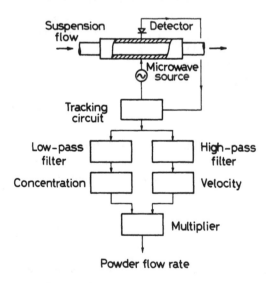

Figure 4.19 Microwave flowmeter.

frequency is detected by a tracking circuit. Then the concentration and velocity are obtained by the lower-frequency component and higher-frequency component of the signal, respectively. The powder flow rate will be obtained as a product of
the concentration and the velocity. The flowmeter was applied
to measure the flow rate of pulverized coal injected into a blast
furnace with mass flow ratio between 10 and 60 [22]. The measured value, however, depends on the dielectric constant of particles, as in the case of the electric capacitance method described
in Section 4.5.3. Therefore, the water content of powder and
fine particles adhered on the inside wall of the resonator will affect the accuracy.

4.6 STATISTICAL METHOD

The velocity measurement in the microwave flowmeter is, as
discussed above, based on the spatial filtering. However, to
utilize the method, the particles have to be shaped into clouds.
Fortunately, the distribution of particles cannot be uniform in
the turbulent flow, as mentioned above. The formation of clouds
is, however, statistical. If a cloud is formed in a flowing gas-
solid suspension, an electric pulse will be observed by an insulated pipe section [23]. The pulse will be utilized in measurement of the particle velocity. Although there may be several
methods for the velocity measurement, the so-called correlation
technique is discussed here.

Figure 4.20 shows an example of a velocity measurement
system. A pair of sensors is set along the conveyor line with a
distance L. The signals detected by the sensors will fluctuate
because of the flow turbulence. That is, the signals contain the
noise components. As discussed in Section 2.2, the cross-correlation will be useful in the analysis of random signals. In actual
application, the cross-correlation is approximated by the following
equation:

$$\phi_c(\tau) = \frac{1}{T} \int_0^T f_1(t)f_2(t + \tau) \, dt \tag{4.6.1}$$

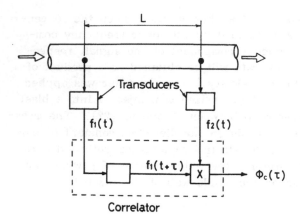

Figure 4.20   Correlation technique for particle velocity measurement.

If the calculation is carried out for various values of the delay-time parameter τ, the cross-correlation takes a maximum at a certain value of $\tau = \tau_0$. The fact shows that there is a correlation between the two signals even if they seem to be independent. Therefore, τ_0 gives the time interval traveling the particle clouds between the upstream sensor and the downstream sensor.   Then the particle flow velocity v is determined by the following equation:

$$v = \frac{L}{\tau_0} \tag{4.6.2}$$

The most important assumption of the correlation technique is that the particle clouds do not disappear during travel.   Even though a longer distance gives higher accuracy in the determination of τ_0, the clouds might disappear.   Therefore, there may be an optimum length L, and it should be determined by experiment. It may also depend on the type of sensor and the scale of clouds. Reported data on this subject are L/D = 1.5 to 2 for an electrostatic sensor [24] and 4 to 6 for a capacitance sensor [25], where D is the pipe diameter.

As is easily recognized, all types of signals will be utilized in the measurement: electric current generated by static electrification of particles, discussed in Section 4.5.1; electric capacitance,

discussed in Section 4.5.3; microwave resonance, discussed in
Section 4.5.5; conductivity for slurries, temperature fluctuation,
density fluctuation, and so on.   Therefore, photometer, ultrasonic density meter, thermocouples, and so on, will also be utilized as their sensors.

The powder flow rate W will be determined as the product of
the particle velocity v (obtained as mentioned above) by particle
concentration c and the cross-sectional area A.   Therefore, the
next problem is to measure the particle concentration.   Although
the problem is discussed in Chapter 5, the statistical method
developed by Beck et al. [25] is discussed here.

The autocorrelation of the capacitance signal $\phi(\tau)$ is given
by Wiener-Khinchin's relation:

$$\phi(\tau) = \frac{1}{2\pi} \int_{-\infty}^{\infty} \Phi(\omega)e^{j\omega\tau}d\omega \qquad (4.6.3)$$

where $\Phi(\omega)$ is the power spectrum of the signal.   The signal
detected by a capacitance electrode is approximated to be a white
noise.   Then the power spectrum $\Phi(\omega)$ for the signal is constant
in all frequency ranges, as shown in Eq. (2.2.8).   Therefore,
Eq. (4.6.3) is simplified by the use of Euler's equation (2.2.7)
and the cutoff frequency of the capacitance transducer ω_0 as
follows;

$$\phi(\tau) = \frac{\Phi}{2\pi} \int_{-\omega_0}^{\omega_0} e^{j\omega\tau}d\omega$$

$$= \frac{\Phi}{\pi} \frac{\sin \omega_0\tau}{\tau} \qquad (4.6.4)$$

From the autocorrelation $\phi(\tau)$, the peak value $\phi(0)$ is given by

$$\phi(0) = \lim_{\tau \to 0} \phi(\tau) = \frac{\Phi}{\pi} \omega_0 \qquad (4.6.5)$$

The cutoff frequency ω_0 depends only on the characteristics of
the transducer.   In other words, ω_0 is independent of particle
concentration.

The power of the noise is represented by the product $\Phi\omega_c$,
where ω_c is the cutoff frequency of the noise.   The cutoff

frequency ω_0 mentioned above is much smaller than the cutoff
frequency of the noise ω_C. It is assumed that the power of the
noise is proportional to the square of the number of particles in
the capacitor. The meaning of the assumption is not clear, but
it may be similar to the case of electric power, which is proportional to the square of current. It is assumed that the cutoff
frequency ω_C is proportional to the number of particles. The
number of particles may correspond to the electric current,
because the movements of particles generate the noise. If the
assumption is satisfied, the power spectrum Φ is proportional to
the number of particles in the capacitor, because the power is
given by the product $\Phi\omega_C$. The number of particles in the capacitor is proportional to the particle concentration if the particle density is constant. Thus the particle concentration is
given by the following equation:

$$c = k\,\phi(0) \qquad\qquad\qquad (4.6.6)$$

where k is the proportional constant. Therefore, the powder
flow rate W is determined by

$$W = \frac{kLA\phi(0)}{\tau_0} \qquad\qquad\qquad (4.6.7)$$

The statistical method is truly obstructionless and is preferable
for the process instrumentation. The calculation is, however,
complicated and a correlator or microcomputer with suitable interfaces is necessary. The method will also be applied to the bulk
solids flow in hoppers or shoots by use of a sound noise pickup
(microphone), because the mutual collision between particles will
produce acoustic emissions [26]. Fiber optics will also be utilized
in order to apply the correlation method to detect the velocity
of particles flowing on a shoot or moving in a vessel.

4.7 OPTICAL METHOD AND OTHERS

There are several methods for the measurement of powder flow
rate, which are not as rigorous as the methods described in

Sections 4.2 to 4.6. The idea, however, may be interesting.
The laser anemometry will also be discussed.

The first idea is to measure the cross-sectional area of powder
bed on a conveyor belt. If the area is measured, the volumetric
flow rate will be obtained as the product of the area and the belt
speed. A part of the belt conveyor is covered by a mask to
make it dark. Then a light irradiates normally to the powder bed,
and the reflected light is detected by a photomultiplier which is
collimated to the illuminated powder bed at some angle. It may
be possible to make the output of the photomultiplier proportional
to the cross-sectional area by adjusting the view angle. The
method will be applied to powders with high reflectance such as
soap powder.

The second idea is based on the volume displacement. A
spiral blade is inserted in a vertical pipe where particles flow
down by gravity. Then the spiral blade will be rotated, and the
number of rotation will be proportional to the volumetric flow
rate of particles. The method is similar to that of the turbine
flowmeter for fluids. The spiral blade suffers from abrasive
wear and might also obstruct the powder flow, resulting in a
blockage.

The third idea is to utilize a pair of level detectors. The
level detectors are set on the wall of a vertical pipe. A flap
valve at the bottom of the pipe supports particles accumulated
in the pipe. Particles gradually accumulate in the pipe, and the
level reaches lower and higher detector positions, successively.
The duration of time between lower and higher positions will be
measured by use of a timer. As a matter of course, the volume
between the two level detectors is known. Then the volumetric
flow rate is given by the volume divided by the duration of time.
If the powder load on the flap valve exceeds a certain limit, the
valve will be opened by the load.

The fourth idea is to utilize the heat removed by a flowing
gas-solid suspension. The metallic powder flowmeter developed
by Union Carbide Co. detects the temperature of gas-solid suspension flowing in a pipeline by the use of a thermistor. Two
thermistors are used in the flowmeter circuit. One thermistor
is mounted in the wall of a Teflon tube supplying the gas, and
another is in the wall of an identical tube conveying the gas-
solid suspension. The electric bridge with two thermistors keeps

the balance when particles are not contained in the flow. If
particles are fed into the flow, heat transfer from the thermistor
set in the wall of the suspension flow line is enhanced by particles
resulting in temperature decrease. Then the electric bridge
loses the balance, and a voltage output will be obtained. The
output can be related to the powder flow rate through calibration.
To prevent temperature variations along the flow line, the flowmeter is enclosed in an insulated case maintained at a constant
temperature by a control circuit.

The following discussion is concerned with the laser anemometry
measurements in a gas-solid suspension flow. The laser light may
penetrate the suspension flow only intermittently with increase
in solids loading, and the mass flow ratio m is somewhat limited
below 2.

The so-called fringe mode is widely used in velocity measurement. The principle is shown schematically in Fig. 4.21a. A
He-Ne gas laser (wavelength 0.6328 μm) is usually utilized to
provide coherent light, whose wave phase is uniform. The beam
is divided into two equal-intensity beams by a beam splitter and
a surface mirror. The two beams are focused to a small region
in the flow by a focusing lens. A particle passing through the
intersection region will scatter the light toward the collecting
optics. Figure 4.21b shows the interference fringes formed at
the intersection region, where the intensity of light is alternately
changed to be dark and bright. The distance between the fringes
is calculated as $\lambda/2 \sin \theta$, where 2θ is the angle of beam intersection and λ the wavelength. Therefore, the intensity of the
scattered light will change with the frequency given by the
following equation:

$$f = \frac{2v_x \sin \theta}{\lambda} \qquad (4.7.1)$$

where v_x is the velocity component, which is normal to the
bisector of the beams.

The collecting optics of the photomultiplier include an aperture
for signal optimization and a pinhole for preventing spurious
signals reaching the photocathode. The resulting frequency of
the signal can be obtained by such methods as spectrum analysis,
frequency counting, or frequency tracking.

(a)

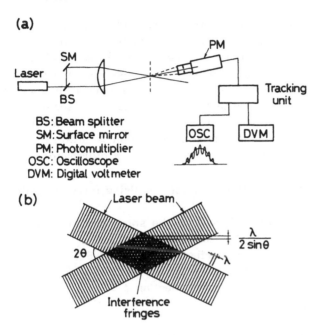

BS: Beam splitter
SM: Surface mirror
PM: Photomultiplier
OSC: Oscilloscope
DVM: Digital voltmeter

(b)

Figure 4.21 Laser anemometry.

There may be some small particle suspended in the flow other than the conveying particles. The two groups of scattering particles, which are different in size and each of which is moving with a different velocity, will produce separate and distinct signals in amplitude and frequency. Therefore, the gas velocity and the particle velocity are obtained by suitable signal processing [27−29].

4.8 FLOW DETECTORS

The so-called flow detector is the powder flowmeter, which only gives data on whether or not the powder flows. The accuracy of measurement of the powder flow rate is not required to be as good as that of the usual flowmeter. The mechanism of flow detectors are also usually simpler than the flowmeters. Some types of flow detectors, called flow switches, sound an alarm when there is no powder flow.

These simple detectors are useful because blockage of pipe-
lines or hoppers will often be caused by particles in powder
handling processes.   Also, there may be damage to or trouble
with belt conveyors, sieving machines, screw feeders, and so on.
    Principles of the detectors are based on the various phenomena
described in this chapter.   Some of the level meters described in
Chapter 6 will also be applied as flow detectors.   The detectors
should not be expensive, and simple construction is preferable.
The following detectors will satisfy the process requirements:

1. An impact flow switch, based on the impulsive force described
    in Section 4.3.
2. An electrostatic detector, based on the statis electrification
    of particles described in Section 4.5.1.
3. A piezoelectric detector, described in Section 4.5.2.
4. A capacitance flow switch, based on the electric capacitance
    increase by particles described in Section 4.5.3.
5. An acoustic detector
6. An optical detector
7. A pressure transducer for gas-solid suspension flow
8. An ultrasonic flow detector
9. A tuning fork detector

    The acoustic detector is a microphone which picks up the
acoustic emission caused by powder flow.   Also, a vibration
detector such as a differential transformer will be utilized.   These
sensors are usually set on the wall of powder handling equipment
with suitable protectors.   Particle flow will damage the sensors
if the setting is not adequate.
    The optical detector is composed of a photo source such as
a luminescent semiconductor diode and a photodetector such as
a photodiode.   Deposition of dust on the optical parts will cause
detection error.   Additional equipment such as an air purge
system will be necessary to keep the optical parts clean.   The
ultrasonic flow detector and tuning fork detector originate from
the powder level sensors and are discussed in Chapter 6.
    Further discussion of flow measurement in pneumatic transport
will be found in Ref. 30, where flowmeters based on electric
capacitance, Coriolis force, and correlation technique are examined.
Reference 31 reviews on-line measurement techniques in coal-
handling systems.

REFERENCES

1. E. G. Parkins and J. B. Alcorn, Process Technol. Int.
    11, 877 (1972).

2. V. Whittaker, Instrum. Technol. 22(6), 45 (1975).

3. G. A. Turner, Trans. Inst. Eng. Aust. 51, 1 (1973).

4. K. Iinoya, T. Yoneda, N. Kimura, K. Watanabe, and T.
    Shimizu, J. Assoc. Powder Technol. 3, 424 (1966).

5. W. Barth, Chem. Ing. Tech. 29, 599 (1957).

6. J. L. Synge and A. Schild, Tensor Calculus, Toronto Univer-
    sity Press, p. 162 (1949).

7. H. Masuda, Y. Ito, and K. Iinoya, J. Chem. Eng. Jpn.
    6, 278 (1973).

8. L. Farber, Trans. ASME 75, 943 (1953).

9. K. Iinoya and K. Goto, Kagaku Kogaku 27, 80 (1963).

10. H. Kagami, M. Maeda, and E. Yagi, Kagaku Kogaku Ronbun-
    shu 1, 327 (1975).

11. H. Masuda and K. Iinoya, AIChE J. 24, 950 (1978).

12. H. Masuda, T. Komatsu, and K. Iinoya, AIChE J. 22, 558
    (1976).

13. H. Masuda, N. Mitsui, and K. Iinoya, Kagaku Kogaku
    Ronbunshu, 3, 508 (1977).

14. P. M. Heertjes, J. Verloop, and R. Willems, Powder Technol.
    4, 38 (1970/1971).

15. T. Patureaux, F. Vergnes, and J. P. Mihe, Powder Technol.
    8, 101 (1973).

16. U. Mann and E. J. Crosby, Ind. Eng. Chem. Process Des.
    Dev. 16, 9 (1977).

17. S. K. Friedlander, Smoke, Dust and Haze, John Wiley &
    Sons, Inc., New York, Chap. 4 (1977).

18. G. Raso, G. Tirabasso, and G. Donsi, Powder Technol.
    34, 151 (1983).

19. Lord Rayleigh, Philos. Mag., Ser. 5 34, 481 (1982).

20. H. Nakajima and T. Tanaka, Kagaku Kogaku, 32, 94 (1968).

21. J. M. Dotson, J. H. Holden, C. B. Seibert, H. P. Simons,
    and L. D. Schmidt, Chem. Eng. 56(10), 128 (1949).

22. S. Kobayashi and S. Miyahara, <u>Keisoku Jidoseigyo Gakkai Ronbunshu</u> <u>20</u>, 529 (1984).

23. H. Masuda, N. Mitsui, and K. Iinoya, <u>Kagaku Kogaku Ronbunshu</u> <u>3</u>, 457 (1977); English translation, <u>Int. Chem. Eng.</u> <u>19</u>, 287 (1979).

24. P. A. Arundel, <u>Chem. Process.</u> <u>18</u>(11), 77 (1972).

25. M. S. Beck, J. Drane, A. Plaskowski, and N. Wainwright, <u>Powder Technol.</u> <u>2</u>, 269 (1968).

26. T. Horiuchi, <u>Keisoku To Seigyo</u> <u>14</u>, 547 (1975).

27. S. Einav and S. L. Lee, <u>Rev. Sci. Instrum.</u> <u>44</u>, 1478 (1973).

28. C. R. Carlson and R. L. Peskin, <u>Int. J. Multiphase Flow</u> <u>2</u>, 67 (1975).

29. A. Birchenough and J. S. Mason, <u>Powder Technol.</u> <u>14</u>, 139 (1976).

30. M. P. Mathur and G. E. Klinzing, <u>Powder Technol.</u> <u>40</u>, 309 (1984).

31. H. H. Kleizen and J. Van Brankel, <u>Powder Technol.</u> <u>40</u>, 113 (1984).

5

Particle Concentration
in Suspensions

Methods for the measurement of particle concentration are discussed
here. Some of the methods were described in Chapter 4 in relation to the measurement of powder flow rate in gas-solid suspension
flow. The concentration measurement of very fine particles
suspended in the air is also discussed in this chapter. In Section
5.1 we outline measurement principles, and in Section 5.2 we
define particle concentration.

5.1 INTRODUCTION

Table 5.1 shows methods for the measurement of relatively high
particle concentration in powder handling processes. These
methods were explained in Chapter 4 in relation to powder flowmeters, except for the gamma-ray densitometer. The methods
listed in Table 5.2 will be applied to fine particles. Dust control
equipment such as fabric filters or electrostatic precipitators will
be monitored by use of measuring instruments based on photo

Table 5.1   Powder Concentration Meter

Principle	Method
Electric capacitance increase	Sec. 4.5.3
Gamma-ray attenuation	Gamma-ray densitometer, Lambert-Beer law
Statistical correlation of various noise	Sec. 4.6
Microwave resonance	Sec. 4.5.5

extinction or electrostatic charge. The optical particle counter
and condensation nuclei counter (CNC) are utilized in the measurement of very fine particles suspended in the process environment—
for example, in a clean room for LSI manufacturing factories.

Table 5.3 shows methods for detecting the mass of fine
particles sampled on a filter or sampled into the sensing zone
of an instrument. The sampling should be done carefully as
discussed in Chapter 3. Particles sampled on a filter will also
be utilized in the laboratory analysis of other properties, such
as particle shape, particle density, or chemical components.

Table 5.2   Dust Concentration Meter

Principle	Method
Photo extinction	Photometer, transmissometer, Lambert-Beer law
Light scattering	Photometer, nephelometer, optical particle counter
Laser radar	Lidar (long-range measurement)
Nucleated condensation	Condensation nuclei counter (ultrafine particles)
Electrostatic charge	Sec. 4.5.1

Table 5.3 Dust Monitor (Sampled Particles)

Principle	Method
Beta-ray attenuation	Beta gauge, Lambert-Beer law
Light attenuation	Lambert-Beer law
Electrically forced vibration	Mechanical resonance (resonant frequency decrease)
Piezoelectric vibration	Piezobalance[a] (resonant frequency decrease)
Electrostatic charging	Charge measurement after unipolar charging

[a]Trade name.

5.2 DEFINITION OF PARTICLE CONCENTRATION

The concentration of particles suspended in air or gas is represented in several ways. The mass flow ratio m defined by Eq. (4.4.1), for example, is utilized in the pneumatic transportation of particles. Holdup particles dw (kg) in a unit volume of gas constitute an important quantity for chemical reaction or mass transfer between gas and particles. The various definitions of particle concentration are summarized as follows:

1. Mass flow ratio, m (unitless):

$$m = \frac{W}{Q} \qquad (4.4.1)$$

2. Flow concentration, c (kg/m^3):

$$c = \frac{W}{Q_v} = \rho_a m \qquad (5.2.1)$$

where Q_v is the volumetric gas flow rate (= Au for pipe flow). If the concentration is divided by the particle density ρ_p, it becomes nondimensional. The dimensionless concentration is called the volumetric concentration.

3. Holdup concentration, c_h (kg/m^3):

$$c_h = \frac{dw}{V} = \frac{W\ d\ell}{vV} = \frac{W}{vA} \tag{5.2.2}$$

where V is the volume occupied by suspension $(= A d\ell$ for pipe flow) and v is the mean velocity of particles. If the holdup concentration c_h is divided by the particle density ρ_p, it becomes nondimensional. The dimensionless concentration is also called the holdup volume concentration, or simply the volumetric concentration.

For simplicity, the three types of concentration mentioned above are often called simply particle concentration. However, the flow concentration c does not coincide with the holdup concentration c_h when the velocity ratio u/v is not unity.

$$c_h = c\ \frac{u}{v} \tag{5.2.3}$$

The particle velocity v is usually smaller than the gas velocity u, and the holdup concentration is higher than the flow concentration. As the holdup concentration represents the mass of the holdup particles in a unit volume of gas at any moment, the higher particle velocity gives a lower concentration for the constant powder flow rate.

The flow concentration c is utilized in the analysis of dust control equipment, because the mass flow rate of particles into or out of the equipment is necessary in order to obtain the dust collection efficiency. If the flow concentration is obtained, the particle mass flow rate will be calculated by use of Eq. (5.2.1) as the product of concentration times the volumetric gas flow rate. However, the velocity of a fine particle [smaller than about 10 μm, depending on the inertia parameter defined by Eq. (4.4.17)] is nearly equal to that of gas, and the concentration is approximated by the holdup concentration c_h. The concentration measured by a photometer or a gamma-ray densitometer is the holdup concentration.

The number concentration is also utilized instead of the mass or volumetric concentration. The optical particle counter, for example, measures the number concentration.

5.3 ELECTRICAL METHOD

The electrical phenomena described in Section 4.5 are also utilized in the measurement of particle concentration. They are the electrostatic charging of particles, piezoelectric effect of crystals, electric capacitance increase of a parallel-plate condenser, and the resonant frequency decrease of microwaves. Also, the induction charging is utilized for electroconductive particles such as metallic powder or liquid droplets. As the subjects were discussed in Section 4.5, only a brief description is given here.

5.3.1 Electrostatic Charging

The electric current generated by particle impact is proportional to the powder flow rate as shown by Eq. (4.5.3). Therefore, the flow concentration c of a suspension will be obtained if the suspension is sampled with constant velocity. Particles in sampled gas collide with the sensor target placed in the sample flow or strike an inside wall of the sensor tube. Any sensitive material can be used for the sensors. The sensors should undergo a test run prior to practical applications, because the surface characteristic of the sensor changes gradually with the number of impacts experienced. The impact efficiency introduced in Eq. (4.5.5) is one of the important factors affecting the sensitivity of the target sensor. This efficiency is also important in the case of sensor tube. A spiral tube increases the sensitivity because of the enhanced impact efficiency. These sensors need calibration, and the calibration curve changes with the particle type.

5.3.2 Piezoelectric Effect

The piezoelectric sensor described in Section 4.5.2 can be applied to the measurement of particle concentration. As mentioned there, the sensor detects the frequency of impacts. Therefore, the particle masses are necessary in order to transform the frequency into the corresponding mass concentration. The concentration given by Eq. (4.5.6) with the local mass flux J is the holdup concentration. That is,

$$c_h = \frac{J}{v} \qquad\qquad (5.3.1)$$

and the flow concentration is given by the equation

$$c = \frac{J}{u} \qquad\qquad (5.3.2)$$

5.3.3 Electric Capacitance Increase

The electric capacitance of a parallel-plate condenser increases with increasing particle concentration, as discussed in Section 4.5.3. The concentration c_V in Eq. (4.5.10) is the holdup volume concentration. Therefore, the mean particle velocity should be known to obtain the powder flow rate or the local mass flux. As the area of the parallel-plate condenser is made as small as 1 cm^2, the sensor is utilized to measure the local concentration of particles in powder handling equipment such as fluidized bed [1].

5.3.4 Microwave Resonance

From Eqs. (4.5.15) and (4.5.16), the holdup volume concentration c_V is given by the following equation:

$$c_V = \frac{2\varepsilon_0 \int |E|^2 \, dV}{(\varepsilon_s - \varepsilon_0)|E|^2 V} \frac{\omega_0 - \omega}{\omega_0} \qquad\qquad (5.3.3)$$

5.3.5 Induction Charging

The electroconductive particles will be polarized in an electric field as shown in Fig. 5.1a. If the polarized particle contacts the target electrode, negative or positive charge will be neutralized by the opposite charge supplied from the electrode, and the particle obtains net charge after the contact. Therefore, the current required for the charging will be found between the electrodes as a voltage pulse [2]. Although the pulse height will be utilized in the measurement of particle size, the number of pulses found in unit time can give the number concentration as in the case of a piezoelectric sensor.

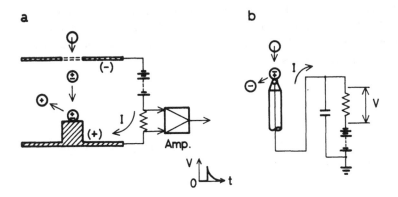

Figure 5.1 Measurement of number concentration by induction charging.

The electrode shown in Fig. 5.1b will also be utilized. There is only one electrode in the figure, but another electrode is the earth. Therefore, the particle will be charged according to its capacitance. The method is also applied to particles of organic liquids such as fuel [3]. Particle size measurement by this method is discussed in Section 7.5.1.

5.3.6 Electrically Forced Vibration (Resonance Technique)

The five methods mentioned above do not require collection of particles by use of a filter or precipitator. The following methods are based on the electrically forced vibration of collected particle mass. Suspended particles are sampled isokinetically as described in Section 3.2.1. The particles are collected on a fibrous filter which is connected to a spring and electronically vibrated. The vibrating motion is represented by the following equation:

$$m \frac{d^2 x}{dt^2} + kx = F \sin \omega t \qquad\qquad (5.3.4)$$

where x is the displacement from the equilibrium position, m the mass of the filter with collected particles, k the spring constant, and F the vibration force.

The first term of Eq. (5.3.4) is the inertia force, and the second is the restoring force of spring. $F \sin \omega t$ is the forced vibration term. Equation (5.3.4) is rewritten with the natural frequency ω_0 as follows:

$$\frac{d^2 x}{dt^2} + \omega_0^2 x = f \sin \omega t \qquad (5.3.5)$$

where $f = F/m$. The natural frequency ω_0 is given by the following equation:

$$\omega_0 = \sqrt{\frac{k}{m}} \qquad (5.3.6)$$

The solution of Eq. (5.3.5) is

$$x = \frac{f \omega}{\omega_0^2 - \omega^2} \left(\frac{\sin \omega t}{\omega} - \frac{\sin \omega_0 t}{\omega_0} \right) \qquad (5.3.7)$$

If the frequency ω of the forced vibration coincides with the natural frequency ω_0, resonance will be found. Therefore, if the resonance frequency is detected, the mass m will be calculated by use of Eq. (5.3.6). As the particles deposit on the filter, the mass m increases from m to m + Δm, and the resonance frequency changes from ω_0 to ω_1. From Eq. (5.3.6), Δm is represented by

$$\Delta m = k \left(\frac{1}{\omega_1^2} - \frac{1}{\omega_0^2} \right) \cong -2 \frac{k}{\omega_0^2} \frac{\Delta \omega}{\omega_0} \qquad (5.3.8)$$

Therefore, if the frequency shift $\Delta \omega$ (rad/s) is measured, a mass of particles will be obtained [4]. Then the flow concentration c is calculated by the following equation:

$$c = \frac{\Delta m}{Q_V \Delta t} \qquad (5.3.9)$$

where Δt is the duration of sampling and Q_V is the sampled flow rate.

5.3.7 Quartz Crystal Microbalance

The mass sensor utilizing a quartz oscillator is based on the principle as discussed above. The sensor is called the QCM (quartz crystal microbalance) [5]. Suspended fine particles are sampled by use of a suction pump and are collected on the quartz crystal. To collect the particles on the surface of the quartz crystal, a small-scale electrostatic precipitator or an impactor (which collects highly inertial particles on a plate) will be utilized. If the particle deposition is tight, the deposited particles increase the mass of the crystal. Therefore, the resonant frequency decreases as discussed above. The frequency shift Δf (Hz) is given by the following equation:

$$\Delta f = -C_f \, \frac{\Delta m}{S} \qquad\qquad (5.3.10)$$

where S is the particle-collection area (m^2) and C_f is the mass sensitivity of the crystal ($Hz \cdot m^2/kg$). The particles are assumed to be collected uniformly on the crystal surface.

The mass sensitivity C_f is represented by use of the natural frequency of the crystal f_0 (Hz), density of the crystal ρ_s (kg/m^3), and the wave propergation velocity in the crystal v_s (m/s) as follows:

$$C_f = \frac{2f_0^2}{\rho_s v_s} \qquad\qquad (5.3.11)$$

The quartz crystal utilized in QCM is a thin plate, as shown in Fig. 5.2. The mode of vibration is selected so that the crystal is most sensitive to the addition or removal of mass. It is possible to make a resonator with all unwanted modes sufficiently suppressed and separated from the appropriate mode. The mode is called the thickness-shear mode, where the two major surfaces of the crystal resonator are always antinodal. To make a quartz crystal plate oscillate in the thickness-shear mode, the plate must be cut to have a specific orientation with respect to the crystal axes. The longitudinal direction of a hexagonal quartz is called the optical axis and the direction of line combining two opposite apices of the hexagon perpendicular to the optical axis is called

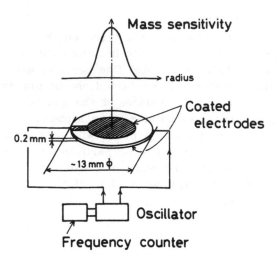

Figure 5.2   Mass sensor (quartz oscillator).

the electrical axis.   The remaining axis, which is perpendicular
to both the optical and the electrical axes, is called the mechanical
axis.   The so called AT-cut crystal (crystal plate cut along the
electrical axis with cutting angle 35°15' from the optical axis)
is one of these plates, and the resonant frequency of the crystal
plate is nearly temperature independent in the neighborhood of
room temperature.

With the density of the quartz (2650 kg/m^3) and the wave
propagation velocity (3340 m/s), Eq. (5.3.10) reads

$$\Delta f = -5.65 \times 10^6 \frac{\Delta m}{S} \qquad\qquad (5.3.12)$$

for a quartz crystal with f_0 = 5 MHz.   The mass sensitivity C_f
(= 5.65 × 10^6) is independent of the physical properties of the
particles deposited.

A frequency shift as small as 0.1 Hz is measured by use of
a simple frequency counter.   Therefore, Eq. (5.3.12) shows that
the QCM can detect small amount of particles of 1.8 × 10^{-8} kg/m^2
(= 1.8 ng/cm^2).   It is however, difficult to measure high concentration suspensions, because the mass sensitivity changes when

the frequency shift exceeds about 10% of the natural frequency f_0. Further, the mass sensitivity shows a Gaussian distribution with a maximum sensitivity at the center of the sensing area. Therefore, calibration is necessary in the actual application of QCM. Also, the adhesive force of particles should be large enough so that the particles would not be dislodged by high-frequency vibrations. The measurement is, in this sense, restricted to fine particles smaller than about 2 μm.

For particle collection by an electrostatic precipitator, the mass sensitivity will decrease when the collected mass is too great, because the particles agglomerate on the crystal surface. The relative humidity of air will also induce two types of errors; one is associated with the crystal and its electrode, and the other stems from the hygroscopic nature of the aerosol deposit. The former is significant only in the case of particle collection by electrostatic precipitation; the latter is substantial in the QCM [6].

Additional reference quartz oscillator will be used to compensate for the change of surrounding conditions, such as temperature and pressure. The reference oscillator is, however, not always effective because airflow patterns around the two oscillators are not always the same. The effect of inlet temperature fluctuations may be eliminated by placing a heat sink (a short piece of thermally conductive tube) in the inlet airstream, because temperature changes are usually of short duration [6].

5.4 ATTENUATION TECHNIQUE

Energy of waves such as light (electromagnetic wave), sound, or radioactive rays will be reduced by suspended particles or a powder bed in their pathway because of wave absorption or scattering. If a light beam passes through a uniformly suspended aerosol layer, a decrease in the intensity of light dI is proportional to the intensity of the incident light I and the thickness of the aerosol layer dL.

$$dI = -\sigma I \, dL \qquad\qquad (5.4.1)$$

where σ is a proportional constant called the extinction coefficient.

From Eq. (5.4.1), the following equation will be obtained:

$$\frac{I}{I_0} = e^{-\sigma L} \qquad\qquad (5.4.2)$$

where I_0 is the initial incident intensity of light ($\cong$ intensity of the transmitted light without particles) and L (m) is the total thickness of the aerosol layer. Equation (5.4.2) shows that the intensity of the light beam exponentially decreases with the thickness of the layer. The fact is known as the Lambert-Beer law.

The proportional constant σ (extinction coefficient) is represented by the following equation:

$$\sigma = \frac{\pi}{4} n \int Q_e f^{(0)} D_p^2 \, dD_p \qquad\qquad (5.4.3)$$

where n is the number concentration of suspended particles, $f^{(0)}$ is the particle size distribution on the number basis, and Q_e is the extinction efficiency (= scattering efficiency + absorption efficiency), which is defined as radiant power scattered and absorbed by a particle divided by radiant power geometrically incident on the particle.

For monodisperse particles, Eq. (5.4.3) reads

$$\sigma = \frac{\pi}{4} n Q_e D_p^2 \qquad\qquad (5.4.4)$$

The number concentration n will be replaced by the holdup concentration c_h by use of the following relationship:

$$c_h = \frac{\pi}{6} D_p^3 \rho_p n \qquad\qquad (5.4.5)$$

Therefore, Eq. (5.4.2) for monodisperse particles becomes

$$\frac{I}{I_0} = \exp\left(-\frac{3 c_h Q_e L}{2 \rho_p D_p}\right) \qquad\qquad (5.4.6)$$

The extinction efficiency Q_e is a function of the particle size and is proportional to D_p^4 for very small particles. Although the

extinction efficiency changes with particle size, it is nearly
constant for particles larger than about 2 μm and converges to
a value of 2 for large particles. The fact is called the extinction
paradox and implies that large particles remove from a light beam
an amount of light corresponding to twice their projected area [7].

The quantity defined by the following equation is called the
mass absorption coefficient η_m (m^2/kg):

$$\eta_m = \frac{3Q_e}{2\rho_p D_p} \tag{5.4.7}$$

By use of the mass absorption coefficient, Eq. (5.4.6) is rewritten
as follows:

$$\frac{I}{I_0} = \exp(-\eta_m c_h L) \tag{5.4.8}$$

Equation (5.4.7) shows that the mass absorption coefficient η_m
is inversely proportional to particle size when the size is much
larger than the wavelength. On the other hand, as Q_e is proportional to D_p^4 when the particle is much smaller than the wavelength, η_m is proportional to D_p^3 for smaller particles. Therefore,
the maximum attenuation will be found for a certain particle size
(between 0.1 and 2 μm). On the other hand, the mass absorption coefficient for very short wavelength such as ultraviolet
(UV) light (0.01 to 0.4 μm) or x-rays (less than 0.01 μm) is
nearly independent of the particle size.

From Eq. (5.4.8), the holdup concentration c_h (kg/m^3) is
given by the following equation:

$$c_h = \frac{1}{\eta_m L} \ln \frac{I_0}{I} \tag{5.4.9}$$

Equation (5.4.9) is applied not only to light but also to radioactive rays and sounds. However, the attenuation of visual
light (wavelength = 0.4 to 0.8 μm) or sound (acoustic method) is
affected by particle size. Sound transmission, especially, is
severely affected by the sound frequency. Therefore, ultraviolet
light or radioactive rays are preferable to a concentration measurement of particles.

The light extinction method is, however, used widely in the
measurement of dust concentrations at the inlet or outlet of
various dust control equipment. In the actual applications, it
is important to avoid dust deposition on the surface of optical
windows facing the suspension flow. The deposited particles
cause the attenuation of light beams, resulting in measurement
error. Figure 5.3 shows an example of the design. Both side
guide tubes are extruded from the duct wall into the aerosol
flow and the tips of guide tubes are beveled. Purging air is
supplied to prevent particle intrusion into the guide tube. An
air purging rate is chosen such that the ratio of air velocity in
the guide tube to local air velocity in the main process flow is
about 20% [8]. If the pressure in the duct is lower than that
in the atmosphere, the air supply holes from the atmosphere on
the guide tube will also be effective.

The attenuation technique is also utilized in the measurement
of a mass of particles in a powder bed or particles collected on
filters. The gamma-ray attenuation is utilized, for example, in
the measurement of powder holdup on a belt scale, described in
Section 4.2. The powder holdup w (kg) is given by the following
equation:

$$w = \frac{A}{\eta_m} \ln \frac{I_0}{I} \qquad\qquad (5.4.10)$$

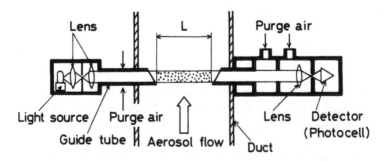

Figure 5.3 Particle concentration meter based on light extinction
(transmissometer).

where I is the gamma-ray intensity penetrating the powder bed which is detected by a scintillation counter, and A is the cross-sectional area of the gamma ray. As the gamma ray can pass through an opaque pipe, it may be utilized as an obstructionless powder concentration meter for gas-solid pipe flow. It is also utilized as a slurry densitometer (concentration meter) with the isotope ^{60}Co (cobalt) or ^{137}Cs (cesium) as a radiation source. For the mass measurement of particles sampled on dust filters, gamma rays are too strong, and beta rays from ^{14}C, ^{147}Pm (promethium), or ^{63}Ni are utilized. The concentration meter based on the beta ray is applicable to measurement of higher concentration than the QCM described in the preceding section. Figure 5.4 shows an example of concentration meter based on beta-ray attenuation. The mass absorption coefficient η_m is a function of the maximum energy E_{max} of the beta ray, and it is nearly independent of the types of materials deposited. Therefore, the meter is easily calibrated by use of a plastic film such as Mylar.

The following equation, for example, gives the mass absorption coefficient [9]:

$$\eta_m = 1.7E_{max}^{-1.43} \qquad (5.4.11)$$

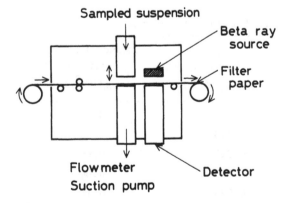

Figure 5.4   Concentration meter based on beta-ray attenuation.

where η_m is expressed in m^2/kg and E_{max} is in MeV (= 1.6 × 10^{-13}J). The constants 1.7 and -1.43 in Eq. (5.4.11) are significantly affected by the geometry of the apparatus, such as the source and detector window thickness, the distance from the source to the detector, and so on. Half-life periods and the maximum energies of the isotopes are listed in Table 5.4. The radioactive rays from isotopes with longer half-lives are stable for longer use of the meter. It is necessary, however, to treat them carefully in applications. Beta-ray sources utilized in dust measurement usually have small intensities (50 to 100 μCi) and are not so dangerous.

5.5 AEROSOL CONCENTRATION

Aerosol is a suspension of solid or liquid particles in a gas, usually air. Aerosols are formed either by the conversion of gases to particles, the disintegration of liquids or solids, or by the resuspension of powdered materials. The term "aerosol" is also used to represent dust, smoke, fume, haze, and mist [10]. Aerosol particle size ranges from 0.001 μm to over 100 μm. Particles formed directly from the gas are usually smaller than 1 μm in diameter.

Table 5.4   Characteristics of Beta-Ray Sources

Source	Half-life (yr)	Maximum energy E_{max}(MeV)	Mass absorption coefficient,[a] η_m(cm^2/mg)
C	5730	0.156	0.24
Co	5.26	0.314	0.09
Ni	92	0.063	0.89
Cs	30	0.514	0.04
Pm	2.62	0.224	0.14

[a]Calculated by Eq. (5.4.11).

Several instruments are used to measure particle size distribution. These instruments are also utilized to measure particle concentration. Photo extinction, beta-ray attenuation, piezoelectric vibration, and others described in the preceding sections are relatively simple methods for the measurement of aerosol concentration. In this section, the light-scattering method and the condensation nuclei counter for ultrafine particles are described.

5.5.1 Light-Scattering Method

Light scattering by an individual particle depends on particle size, the refractive index of the particle, the wavelength of light, and the scattering angle. The problem is discussed in Section 7.4. The method discussed here is based on the measurement of light scattered from many particles suspended in a detection volume, as shown in Fig. 5.5. Aerosol is sampled in a dark detection chamber at a constant flow rate by use of a suction pump. Light scattered by aerosol particles is detected by a photomultiplier or a photodiode and transformed into an electric current. The intensity of detected light is given by a summation of light scattered by the individual particles. The scattering

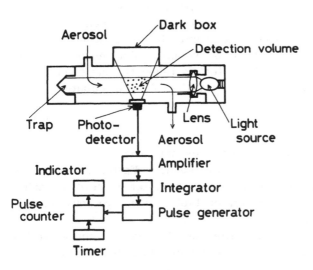

Figure 5.5   Scattering photometer.

angle changes over a wide range of angles depending on the
position of the particle in the detection volume. The so-called
integrating nephelometer is a photometer of this type which is
designed to measure the light scattered over as wide a range of
angles as possible. The instrument can be utilized to measure
the scattering coefficient of aerosols. If the absorption coefficient
is assumed to be zero, the extinction coefficient is measured by
the nephelometer because the extinction coefficient is given by
a summation of scattering coefficient and absorption coefficient.

As mentioned above, the photometer measures the overall
scattering coefficient of aerosols, which will be proportional to
the number concentration as in the case of the extinction coefficient described in Section 5.4. It also depends on the particle
size. The sensitivity is, however, higher than that of the
extinction photometer, and a concentration as low as 3×10^{-6}
g/m^3 is measured by a scattering photometer. The electric
current from the photodetector is amplified and accumulated in
a condenser. When the voltage of the condenser is increased to
a certain value, the condenser will be discharged, resulting in
an electric pulse. The number of pulses per unit time is proportional to the aerosol concentration. The meter is, for example,
calibrated so that the one pulse corresponds to the mass concentration of 10^{-5} g/m^3. Stearic acid particles are often used in
the calibration, but the calibration constant K depends on the
type of aerosol. The electric current will be observed even if
no particles exist in the detection volume. The number of pulses
per unit time caused by this effect is called the dark count D.
Then the concentration is given by the following equation:

$$c_h = K(N - D) \qquad\qquad\qquad (5.5.1)$$

where N is the number of pulses per unit time.

The number of pulses is indicated as a digital number by use
of light-emitting diodes (LEDs), and the meter is sometimes called
a digital dust counter. Small, portable meters are commercially
available in Japan.

There may be multiple scattering when aerosol concentration
is high. The effect depends on the following dimensionless
quantity:

$$\kappa = \int \sigma \, dL \qquad\qquad\qquad\qquad (5.5.2)$$

where σ is the extinction coefficient. The quantity κ is called the optical thickness or depth. For $\kappa < 0.1$, the assumption of single scattering is acceptable, whereas for $0.1 < \kappa < 0.3$, it may be necessary to correct for double scattering. When κ is larger than 0.3, multiple scattering must be taken into account.

   The following discussion is concerned with LIDAR (light detection and ranging) [11,12]. This is also called optical radar, and is a useful tool in pollution monitoring, meteorology, and atmospheric physics. When range information is required, pulsed lasers can be used in a way which is directly analogous to radar. The range and direction of a target are determined from the bearings and the time interval between the transmission of a pulse and the detection of the return signal. It is usual to mount the transmitter and receiver adjacently and detect the back-scattered component of the signal. When the transmitted laser pulse passes through an increased concentration of scattering particles, such as in a smoke plume or cloud, this gives rise to an increase in the scattering signal. The LIDAR have been used to range targets at distances of tens of kilometers.

   It is possible to determine the extinction coefficient of a smoke plume. Several plumes in certain areas are also detected, even if the near plumes attenuate the signals from the more distant targets. It is also possible to monitor the buildup and dispersal of pollution with time. Gaseous pollutants can be identified by the use of absorption spectra. Continuous-wave lasers such as carbon-monoxide systems emit a range of frequencies in the infrared which correspond to absorption lines of some gaseous pollutants.

5.5.2 Condensation Nuclei Counter

The concentration of ultrafine particles can be measured by use of a photometer as described in Section 5.4 or 5.5.1, if the particles grow to micrometer size. Condensation nuclei counters (CNCs) saturate an aerosol with water or alcohol vapor and then cool it by adiabatic expansion or other mechanisms to create the required supersaturation for particle growth.

When dn (mol) of liquid vapor condenses on a nucleus as shown in Fig. 5.6, the free energy change dG/dn is given by thermodynamics as follows [13]:

$$\frac{dG}{dn} = RT \int_{p_s}^{p} \frac{dp}{p} = RT \ln \frac{p}{p_s} \qquad (5.5.3)$$

where p is the partial pressure of vapor on the nucleus, p_s the partial pressure for a flat liquid surface at a given temperature T, and R the gas constant (= 8.31 J·mol^{-1} K^{-1}). The ratio p/p_s is called the saturation ratio.

On the other hand, the surface energy of the condensed droplet increases as follows:

$$\begin{aligned} dG &= \gamma \ dA \\ &= \gamma \ [\pi(D_p + dD_p)^2 - \pi D_p^2] \\ &\cong 2 \ \pi\gamma D_p dD_p \end{aligned} \qquad (5.5.4)$$

where γ is the surface tension of the liquid and A is the surface area. The diameter increase dD_p is given by the following relation:

$$\frac{1}{2} \pi D_p^2 \ dD_p = \frac{M}{\rho} \ dn \qquad (5.5.5)$$

where M is the molecular weight and ρ is the density of the liquid.

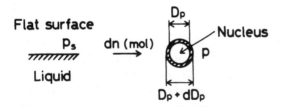

Figure 5.6   Condensation of liquid vapor on a nucleus.

From Eqs. (5.5.4) and (5.5.5), the surface energy increase is given by

$$dG = \frac{4\gamma M}{D_p \rho} \, dn \qquad\qquad (5.5.6)$$

From Eqs. (5.5.3) and (5.5.6), the following equation will be obtained at equilibrium:

$$\frac{p}{p_s} = \exp\left(\frac{4\gamma M}{\rho RT D_p}\right) \qquad\qquad (5.5.7)$$

Equation (5.5.7) is called the Kelvin or Thomson-Gibbs equation, and the diameter determined by the equation is called the Kelvin diameter. The equation also shows that the partial pressure of liquid vapor at the surface of a small droplet is greater than the saturation vapor pressure. This effect is called the Kelvin effect and is significant only for particles less than 0.1 μm.

When all the nuclei are exposed to the same supersaturation for the same length of time, they grow to about 10 μm in diameter regardless of their initial size. The smallest particles detected by CNC are given by the above-mentioned Kelvin diameter for the supersaturation achieved. Supersaturations of 200 to 400% corresponds to a theoretical lower limit of size detectability of about 0.002 μm (= 20 Å). However, counting efficiency will decrease for smaller particles. On the other hand, if the supersaturation is too high, homogeneous nucleation (self-nucleation) will be caused. Then many droplets will be produced even if no particles exist in the sample gas. Light scattering by a single particle is also utilized in order to count the grown droplets. Instruments that can measure concentrations of 10^4 to $10^{13}/m^3$ are commercially available [14].

REFERENCES

1. P. J. Bakker and P. M. Heertjes, Brt. Chem. Eng. 3, 240 (1958).

2. D. P. Keily and S. G. Millen, J. Meteorol. 17, 349 (1960).

3. K. Tamano, Kagaku Kogaku Ronbunshu 8, 470 (1982).

4. J. C. F. Wang, B. F. Kee, D. W. Linkins, and E. Q. Lynch, Powder Technol. 40, 343 (1984).

5. C. Lu and A. W. Czanderna, Applications of Piezoelectric Quartz Crystal Microbalances, Elsevier Science Publishing Co., Inc., New York (1984).

6. P. S. Daley and D. A. Lundgren, Am. Ind. Hyg. Assoc. J. 36, 518 (1975).

7. W. C. Hinds, Aerosol Technology, John Wiley & Sons, Inc., New York, Chap. 16 (1982).

8. D. L. Fenton and J. J. Stukel, Int. J. Multiphase Flow 3, 141 (1976).

9. G. J. Sem and J. A. Borgos, Staub Reinhalt. Luft 35, 5 (1975).

10. S. K. Friedlander, Smoke, Dust and Haze, John Wiley & Sons, Inc., New York (1977).

11. J. S. Freeman, Process Eng., p. 97 (Apr. 1975).

12. E. E. Uthe, B. M. Morley, and N. B. Nielsen, Appl. Opt. 21, 460 (1982).

13. F. W. Sears and G. L. Salinger, Thermodynamics, Kinetic Theory, and Statistical Thermodynamics, 3rd ed., Addison-Wesley Publishing Co., Inc., Reading, Mass., Chap. 8 (1975).

14. W. C. Hinds, Aerosol Technology, John Wiley & Sons, Inc., New York, Chap. 13 (1982).

6

Level of Powder
in Storage Vessels

Methods for the measurement of powder level are described here
by classifying them into several groups: mechanical, electrical,
ultrasonic wave, radiometric, pneumatic, and other methods.   In
Section 6.1 we outline the measurement principles.

6.1   INTRODUCTION

Level meters are classified into two types [1,2]: continuous and
discrete.   Continuous level meters can be utilized to measure the
distance between the upper surface of a powder bed and, for
example, the top of a storage vessel.   On the other hand, discrete
level meters detect the existing powder bed only at their set
positions.   Discrete level meters are also called level switches and
are usually used to maintain the powder level between predeter-
mined upper and lower limits by controlling the feed or the
discharge amount of powder.

   Table 6.1 shows the principles of continuous level meters.
The electric capacitance increase described in Section 4.5.3 is
also utilized in measurement of the powder level.   Table 6.2
shows the principles of various level switches.   These level
switches are more widely used than are continuous level meters.
   Level meters or level switches should be located at suitable
positions, because a powder bed does not have a horizontal
surface.   Further, the profile of the surface will gradually change
during feeding or discharging processes.   For free-flowing
particles, the profile may be estimated by use of the angle of
repose in general.   If estimation is difficult, experiments should
be carried out to determine a suitable location for the level
meters.   A change in the surface profile during the discharging
process is also affected by the design of storage vessels, and
so-called mass-flow hoppers are preferable.

6.2 MECHANICAL METHOD

Mechanical level meters are long established and most reliable.
Among these level meters, only the sounding level meter described
below (Section 6.2.1) is continuous in type; the others are all
level switches.   They are simple and, usually, not expensive.
Various constructions other than those described here may also
be possible.   For some of these level switches, mechanical bearings

Table 6.1   Powder Level Meter (Continuous Type)

Principle	Method
Weighing	Load cell
Electric capacitance	
Electric resistance	
Sounding	Wire rope with a sinker
Attenuation	Radiation, ultrasonic wave
Reflection time (reflectmetry)	Ultrasonic wave, electromagnetic pulse

Table 6.2 Powder Level Switch

Principle[a]	Method
Back-pressure increase	Gas purge or gas injection
Elastic deformation (powder pressure)	Diaphragm with mercury switch
Mechanical blockage of motion by powder	Reciprocation, rotation, swinging motion
Electrically forced vibration	Piezoelectric vibration, tuning fork
Attenuation	Microwave, electromagnetic wave

[a]Some of the principles shown in Table 6.1 are also utilized in level switches.

may suffer from problems caused by fine particle intrusion. If the motor is overloaded by additional force caused by the fine particles, level switches will give erroneous signals. Therefore, the mechanical bearings should be isolated as in the case of swing-type level switches (Section 6.2.4), although the diaphragms of level switches may suffer particle abrasion.

6.2.1 Sounding Level Meter

Sounding level meters are mounted at the top of storage vessels as shown in Fig. 6.1. A sinker connected by a wire rope is lowered slowly from a wire drum of the level meter by use of a motor. When the sinker reaches the surface of the powder bed, the wire rope is allowed to slacken. Then the torque acting on the wire drum decreases. The change in torque is detected mechanically and the wire rope is rolled up by reversing the motor's rotation. The process is repeated at certain time intervals. The powder bed level is measured according to the length of rope released, which will be determined from the number of rotations of the drum.

Level meters of this type can measure up to 50 m in depth from the top with 0.1 m accuracy. This meter is applicable to

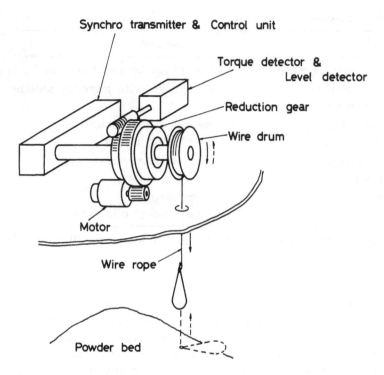

Figure 6.1   Sounding level meter.

a high-temperature vessel as a blast furnace.   The level of
solid materials immersed in liquid can also be measured.   The
motor, however, may become overloaded when the sinker is
buried in a powder bed.   A sinker with suitable shape and
weight should be selected according to the powder properties.
Time intervals of measurements are usually changeable between
1 and 30 minutes.

6.2.2   Piston Level Switch

Figure 6.2 shows a schematic diagram of a piston-type level
switch.   A shaft having a disk at the end is moved reciprocally
by use of a motor and a crank.   A spring and a link are inserted
between the shaft and the motor as shown in Fig. 6.2.

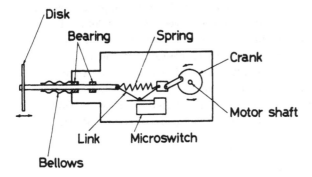

Figure 6.2 Piston level switch.

If the disk is blocked by particles from the left-hand side in the figure, the spring is compressed by motor action and the link changes into a V shape. Then the link pushes a microswitch and the motor is switched off. Also, an electrical signal showing the powder level is produced. If the disk is released from the particles, the link is returned to its original shape by spring action. Then the motor rotates again.

Bearings are isolated by bellows as shown in Fig. 6.2, and dust will not intrude into the mechanical part. However, adhesion of particles on the disk, bellows, and shaft may cause trouble. Piston level switches are not suitable for pressurized vessels, because additional force caused by the pressure difference will act on the shaft.

6.2.3 Rotating Blade Level Switch

A level switch of this type has a blade at the end of a shaft, which is rotated slowly (1 to 3 rpm) by a synchronous motor. Mechanical blockage of the rotating blade by particles is effectively detected by use of a spring and two microswitches. If the blade is blocked by particles, the motor housing rotates in the counter-direction and pushes a microswitch. Then the motor stops and an electrical signal showing the powder level is produced.

If the level of powder is decreased by discharging powder, the blade of the level switch becomes free of particles. Then the rotation of the motor housing is reversed by spring action and the other microswitch is pushed, restarting the motor.

Rotating blade level switches are widely used because of their simplicity. The blade may, however, be broken by firm blockage by powder. If the blade is buried in powder flow, it may be forced to rotate, resulting in an erroneous signal.

6.2.4 Swing Level Switch

Figure 6.3 shows a swing-type level switch. A blade is swung from side to side by a motor and mechanical connectors. Leaf springs are inserted between the blade shaft and the motor. Mechanical parts are isolated from particles by use of a diaphragm. If the blade is blocked by particles, leaf springs will bend. At the same time, the motor, instead of the blade, is moved along a guide, because the blade cannot move. Then the motor pushes a microswitch as in the case of the two types of level switches discussed above. If the blade is released from particles, the motor is returned to its normal position by spring action.

The blade is set vertically in a storage vessel. Therefore, drag force caused by powder flow is not as large as in the case of rotating blade level switches. Thus the erroneous signal

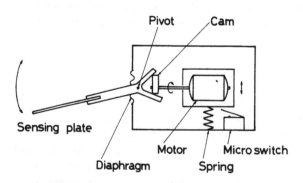

Figure 6.3   Swing level switch.

caused by powder flow mentioned above can be avoided in swing level switches. It is also rarer to suffer damage from firm blockage by powder.

6.2.5 Diaphragm Level Switch

Level switches of this type have no motors. A diaphragm is connected with a microswitch or a mercury switch as shown in Fig. 6.4. If powder pressure acts on the level switch, the diaphragm is elastically deformed and turns the switch on. After the powder level goes down, the diaphragm returns to its normal position and the switch goes off. The sensitivity of the level switch to powder pressure is adjusted by use of a spring (Fig. 6.4). Diaphragms are made from synthetic rubber or metals such as stainless steel.

This type of level switch has no parts protruding into storage vessels, in contrast to other level switches mentioned above. Diaphragm level switches may be susceptible to mechanical damage in this sense.

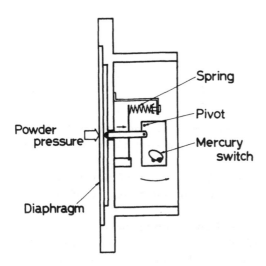

Figure 6.4 Diaphragm level switch.

6.3 ELECTRICAL METHOD

Several electrical properties of powder are also utilized in the
measurement of powder level: electric capacitance, discussed
in Section 4.5.3, and electric resistance between electrodes.
Reflection of electromagnetic pulses by the powder bed surface
and electrically forced vibration, discussed in Section 5.3.6,
are also utilized. Although it is not discussed here, microwave
attenuation through a powder bed may also be utilized as one
of the electrical methods for level switches. Microwave level
switches need no physical contact with particles and will not be
affected by dust, particle adhesion, sound noise, temperature,
and pressure.

6.3.1 Electric Capacitance Level Meter

As discussed in Section 4.5.3, electric capacitance is increased
by particles existing between the electrodes. For continuous
measurement of powder level, a wire electrode is stretched
vertically in a storage vessel. The counterpart electrode is a
vessel itself. For a concentric configuration of the electrode
and a cylindrical vessel, electric capacitance C will be given by
the following equation:

$$C = \frac{2\pi\varepsilon L}{\ln(b/a)} + \frac{2\pi\varepsilon_0(L_0 - L)}{\ln(b/a)}$$

$$= \frac{2\pi\varepsilon_0 L_0}{\ln(b/a)}\left[1 + (\bar{\varepsilon} - 1)\frac{L}{L_0}\right] \qquad (6.3.1)$$

where L_0 is the length of the electrode, L the powder level, a
the diameter of the electrode, b the diameter of the vessel, and
$\bar{\varepsilon}$ (= $\varepsilon/\varepsilon_0$) is a specific dielectric constant. In actual applications
the capacitance will be affected by the shape of the vessels, the
end effect of the capacitor, and the packing density of the
particles.

Electric capacitance level switches, which have a rod or a
bar electrode, are also available. Some of them have an earth
electrode at their base. Level switches of this type have no
moving parts and are easily adapted to high-pressure or high-

temperature conditions. It is, however, necessary to adjust their sensitivity at the installation site, because the sensitivity depends on the shape of the storage vessels. The sensitivity also depends on the dielectric constant of powder materials. Therefore, readjustment is necessary when the type of powder or moisture content of the powder changes. Fine powders adhering to the electrodes may increase the electric capacitance and cause erroneous signals. Electric capacitance level meters or switches cannot be applied to electrically conductive powders.

6.3.2 Electric Resistance Level Meter

A rod-like electrode is inserted in a storage vessel, and constant voltage is applied between the electrode and the vessel wall which works as the counterpart electrode. As the specific resistance of particles is smaller than that of air, electric current through the powder layer becomes larger as the powder level increases. The current also depends on the shape of the vessel. For a concentric configuration of the electrode and a cylindrical vessel, the current I is given by the following equation:

$$I = \frac{V}{\rho_d \int_a^b (dr/2\pi rL)}$$

$$= \frac{2\pi V}{\rho_d \ln(b/a)} L \qquad\qquad (6.3.2)$$

where V is the applied voltage (V), L the level of powder (m), ρ_d the specific resistance of the powder ($\Omega \cdot m$), a the diameter of the electrode (m), and b the diameter of the vessel (m).

The specific resistance depends on the packing density, type of material, moisture content, and temperature. Therefore, the sensitivity is affected by these variables.

6.3.3 Reflection of Electromagnetic Pulse

An electrode such as a copper wire is inserted in a storage vessel as shown in Fig. 6.5. This electrode, together with the grounded vessel, constitutes a transmission line for electromagnetic pulses.

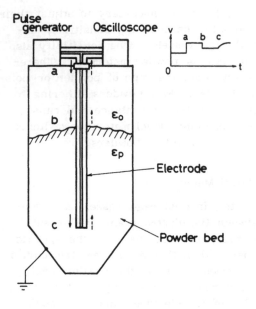

Figure 6.5   Electromagnetic pulse level meter.

An electromagnetic pulse travels along the electrode and is partially reflected when it encounters a discontinuity in the transmission line.   Reflections occur at all points of discontinuity, such as connection parts between the transmission line and the vessel, the interface between air and the powder bed, or the termination of the electrode.   The powder level is obtained by measuring the transit time of the electromagnetic pulse as it travels down the electrode and is reflected by the surface of the powder bed.   The propagation velocity of the pulse is proportional to the inverse square root of the dielectric constant of materials constituting the transmission line.   As the dielectric constant of the powder bed is several times that of air, an adequate discontinuity is obtained at the powder bed surface [3].

6.3.4 Electrically Forced Vibration

Two piezoelectric crystals 1 and 2 are used in this type of level
switch, as shown in Fig. 6.6. A weak vibration of a rod or
tuning fork deforms piezoelectric crystal 1.   Deformation of the
crystal gives an electric output as discussed in Section 4.5.2.
The output is amplified and is supplied to piezoelectric crystal 2
as its input.   Then crystal 2 is electrically excited and finally
vibrates at the resonance frequency. As discussed in Sections
5.3.6 and 5.3.7, the resonance frequency decreases when the
rod or tuning fork is constrained by particles.   The decrease in
vibration frequency is detected by piezoelectric crystal 1.   As
the output of crystal 1 is supplied to crystal 2, the vibration of
crystal 2 is also suppressed.   Thus the level of powder is
detected.

This type of level switch is not affected by the electrical
properties of particles.   The level of powder immersed in liquid
can also be detected.   However, adhesion of fine particles to the
rod or tuning fork causes a decrease in frequency, resulting in
erroneous signals.

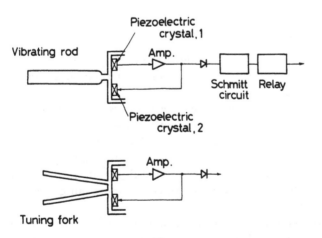

Figure 6.6 Level switches based on electrically forced vibration.

6.4 ULTRASONIC WAVE LEVEL METER

An ultrasonic wave is a sound wave with a frequency higher
than 20 kHz. It has high directivity, and a 50-m distance can
be detected with 1 to 2% accuracy by ultrasonic level meters.
If the ultrasonic wave is emitted from the level meter to a powder
bed, a part of the wave is reflected at the surface of the powder
bed and detected by a sensor of the level meter. The reflection
time of a pulsed ultrasonic wave is utilized to measure the powder
level. The velocity of the wave in air is given by the following
equation:

$$u = 331 + 0.6T \qquad\qquad (6.4.1)$$

where u is the sound velocity (m/s) and T is the temperature
of the air ($^{\circ}$C). The distance from the sensor to the surface
of a powder bed is given by the product of half of the reflection
time and the sound velocity.

Reflection of the ultrasonic wave is shown in Fig. 6.7. The
reflection factor of sound pressure is represented by the following equation:

$$R = \frac{\rho_2 u_2 \cos \theta_i - \rho_1 u_1 \cos \theta_p}{\rho_2 u_2 \cos \theta_i + \rho_1 u_1 \cos \theta_p} \qquad\qquad (6.4.2)$$

where ρ is the density, θ_i the incident angle, θ_p the angle of
refraction, and the subscripts 1 and 2 denote substances through
which the sound passes.

Equation (6.4.2) shows that the ultrasonic wave emitted
vertically to the water surface is perfectly reflected (R = 1),
because the $\rho_2 u_2$ of water is much larger than that of air ($\rho_1 u_1$).
The reflection factor also depends on the packing density of
powder, and it may be much smaller than unity. It is, however,
possible to detect the weak reflected wave by increasing the
sensitivity of a detector.

Figure 6.8 shows an example of an electric circuit for ultrasonic wave level meters. Emission and detection of the wave
are carried out by use of a single element such as a piezoelectric vibrator, an electrostrictive vibrator, or a magnetostric-

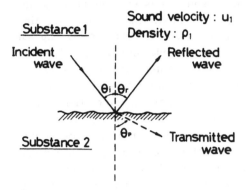

Figure 6.7   Reflection of ultrasonic wave.

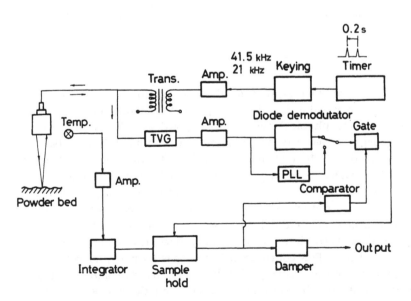

Figure 6.8   Electric circuit for ultrasonic level meters.

tive vibrator. Such an element (sensor) is excited periodically
by a pulsed electric power supply. Then the surrounding air
is vibrated and an ultrasonic pulse wave is detected by the same
element and is transformed into an electric signal. A TVG (time
variation of gain) circuit changes the sensitivity of receiver
according to the reflection time, and a higher sensitivity is
given for a longer reflection time in order to detect the weak
signal. The TVG circuit also eliminates noise in the signal.

The noise-eliminated signal passes through a diode demodulator
or a PLL (phase-locked loop) detector and is shaped into pulse
series. An interval of pulses corresponds to the reflection time
of the ultrasonic wave. The velocity of sound is calculated by
Eq. (6.4.1) and is integrated over half of the time interval.
Thus the level of the powder bed is estimated.

Ultrasonic level meters need no physical contact with the
powder. They are also easy to install. Their sensitivity is,
however, affected by suspended dust or water vapor in storage
vessels because of the sound attenuation. Other sound sources,
such as powder mills, produce noises and may cause errors in
measurement. These factors affecting level meters should be
investigated prior to practical applications of these meters.

6.5 RADIOMETRIC METHOD

The attenuation technique described in Section 5.4 is also used
to detect the powder bed level [3]. Level meters of this type
do not require physical contact with powder. Radiation from
high-energy gamma-ray sources such as cesium 137 or cobalt 60
may be used. Cobalt 60 has a half-life of 5.5 years, and cesium
137 of 33 years. For large vessels, cobalt 60 is suitable.
Cesium 137 is applied to small vessels or low-density materials.
In most cases, detectors are placed opposite the radiation source
to sense a change in the intensity of transmitted radiation.

A Geiger-tube detector is usually used for simple on-off
level switches. For continuous level measurement, more stable
scintillation counters are preferable. Figure 6.9 shows various
arrangements for the radiation sources and detectors. Figure
6.9a shows a level switch. One point source shown in Fig. 6.9b

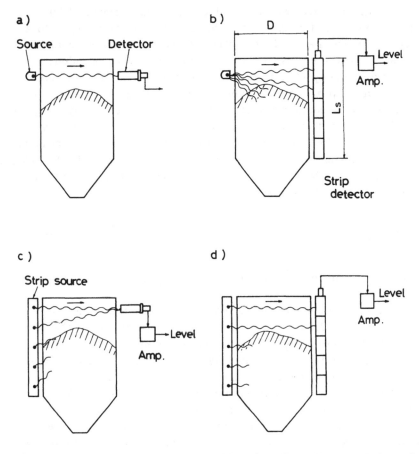

Figure 6.9 Level switch (a) and level meters (b, c, d) based on the radiation method.

provides a nonlinear output because the quantity of radiation shielded by powder bed changes nonlinearly with powder level. Usually, if the vessel diameter D is less than twice the level span L_s, excessive nonlinearity will result.   The nonlinearity can be reduced by the use of multiple point sources. Strip sources and detectors are preferable and can be utilized for high-precision measurements.

Continuous level meters need dc or ac amplifiers. The former needs readjustment of electronic zero drift every 1 to 4 weeks. Therefore, if little instrument maintenance is available in plant operations, use of an ac amplifier is recommended.

Random decay of a radioisotope may cause short fluctuation periods in output signals. Therefore, the output signals will be averaged for a certain period (which can be selected by varying the time constant of detection) to avoid the error caused by fluctuations. The intensity of noise caused by random decay is larger for a shorter time constant of detection. Standard deviation s of the noise intensity is inversely proportional to the square root of the time constant:

$$s = \frac{k}{\sqrt{T_s}} \qquad\qquad (6.5.1)$$

where k is a proportional constant and T_s is the time constant of detection.

The fact will easily be recognized, as the total number of data n in Eq. (2.2.1) is, in this case, proportional to the time constant T_s. Typically, a time constant of about 10 s is used. If the time constant is reduced according to a process requirement, the noise will increase accroding to Eq. (6.5.1).

Usually, a larger radioisotope source is necessary for larger vessels or vessels with heavier walls. Special windows for sources and detectors in the wall with dead space will greatly reduce source size. However, the alignment of the windows is critical. The use of these level meters on empty vessels must be carefully considered to ensure that operating and maintenance personnel will not receive excessive radiation. Human tolerance for safe exposure is about 100 mrem/week. A reflectance level switch should be considered instead of the more common transmission unit for large-scale vessels. Reflectance level switches do not require as large a source as does the transmission type. Although reflectance level switches have some restrictions [4], the installation will be less expensive because of the relative smallness of sources. As an integral-package reflectance unit can easily be driven by a motor, it may also work as a continuous level meter.

When several adjacent vessels are equipped with radiometric
level meters, the orientation of the source beams must depend
on the arrangement of these vessels. When two vessels are
adjacent, the two sources should be placed facing in opposite
directions to avoid interaction between them.

6.6 PNEUMATIC METHOD AND OTHERS

The pneumatic and weighing methods are described in this
section. The optical method for the measurement of the surface
profile of a powder bed is also introduced.

6.6.1 Pneumatic Method

Pneumatic level switches utilize air (or gas) nozzles. Air is
injected into storage vessels through nozzels equipped at side
walls. The air flow is held constant by use of a flow regulator.
If the air jet is blocked by the powder bed, the back pressure
in an air nozzle increases. This pressure can be detected by
the use of a pressure transducer, described in Section 9.3. The
principle of pneumatic level switches is similar to that of the
gas-purge flowmeter described in Section 4.4 or the nozzle flapper
utilized in pneumatic controllers. Air must be continuously
injected even if level measurements are not required, because
the nozzles may be blocked by particle intrusion. For adhesive
particles, a cave may be formed in the powder bed after the
particles near the nozzle are blown off. Then no pressure
increase can be obtained and the level switch fails to detect the
powder bed because of the cave. Even in such a case, if the
powder bed is moving down, the cave will continuously be filled
with powder and the pneumatic level switch will work normally.

6.6.2 Weighing Method

If the purpose of level measurements is to determine the content
of powder in a storage vessel, a weighing method is preferable.
Load cells utilized in hopper scales (Section 4.2) are also applied
in weighing level meters as their sensors. Further description
of load cells is given in Section 9.3. To estimate the level of

powder bed according to the weighing method, numerical calculation must be done based on the bulk density of powder and the
geometrical shape of the vessel. The estimated level is, however,
not always correct, because the bulk density may change because
of the compression of powder during the storage. Powder bridging or rat holing in the storage vessel may also cause estimation
errors.

6.6.3 Optical Method

The optical method has the disadvantage that the light intensity
will be attenuated by particles suspended in the path (see also
Section 5.4). If this problem can be overcome, the method will
be utilized in more complicated measurements because the light
path is easily changed by use of mirrors, prisms, and optical
fibers. A high-powered laser is used to obtain the surface profile of a powder bed [5]. The method developed by Kawasaki
Steel Corp. (Japan) utilizes an argon laser (0.514 μm, 4 W).
The laser beam scans the surface of the powder bed and the
trajectory of light on the surface is recorded by use of a highly
sensitive TV camera, as shown in Fig. 6.10. The data obtained
are analyzed by a microcomputer based on triangulation. As
the distance between the laser scanner and the TV camera is
known, in addition to their azimuths, the triangle is uniquely

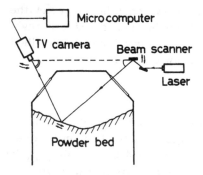

Figure 6.10 Optical measurement of surface profile.

determined. Therefore, the spatial position of the laser spot
can be determined through calculation. The surface profile can
be estimated by the optical method within an error of ±50 mm.
The time needed for measurement is about 10 s.

A pulsed YAG laser (yttrium-aluminum-garnet laser; 1.06 μm,
0.1 J/pulse, and peak value of 10 MW) is used in a similar method
developed by Nippon Steel Corp. (Japan). Fiber optics and an
array of photomultipliers are utilized to detect the light trajectories
with high sensitivity. Measuring systems are controlled by a
computer and the surface profile can be measured within an error
of ±30 mm.

REFERENCES

1. K. Iinoya and H. Masuda, Funryuutai Process No Jidoka (On
   the automation of particulate processes), Nikkan Kogyo,
   Tokyo, p. 65 (1975).
2. K. Iinoya (ed.), Funryuutai Keisoku Handbook (Handbook
   for the Instrumentation of Powder Handling Processes),
   Nikkan Kogyo, Tokyo, p. 341 (1981).
3. W. Fuchs and P. M. Yavorsky, Chem. Eng. 81(13), 145 (1974).
4. S. Rowe and H. L. Cook, Jr., Chem. Eng. 76(2), 159 (1969).
5. H. Yamasaki (ed.), Jido Seigyo Handbook—Kiki Ouyou Hen
   (Handbook for Automatic Control—Instruments & Applications)
   Ohmsha, Ltd., Tokyo, p. 842 (1983).

7

Particle Size

Methods for the on-line measurement of particle size are discussed here by classifying them into several groups: direct classification, permeability, optical, electrical, acoustic, and other methods. In Section 7.1 we outline the principles of the measurement.

7.1 INTRODUCTION

Various principles for on-line measurement are listed in Table 7.1. They are classified as direct or indirect methods. The direct method is based on the particle size classification technique, and indirect methods utilize various phenomena, such as the pressure drop of air flowing through a powder bed. It is generally difficult or expensive to obtain the particle size distribution. Therefore, widely used instruments detect only the mean particle size discussed in Section 1.4, which is the most important parameter in industrial size control. Some instruments are also available to detect particle size distributions. The laser

Table 7.1   Particle Sizer

Principle	Method
Direct classification	Terbo-Pow-Sizer,[a] elutriator sizer, centrifugal sizer, Sieve-Pow-Sizer[a]
Air permeability	Automatic Blaine meter, automatic permeability meter
Laser diffraction	Laser particle sizer
Laser Doppler velocimeter	Simultaneous measurement of particle velocity and size
Holography	Simultaneous measurement of particle size, velocity, concentration, and trajectory
Correlation technique	Optical method (laser particle sizer, fiber optics), electric conductivity method
Induction charging	Simultaneous measurement of number concentration and size
Corona charging	Charging specific surface area meter
Acoustic emission	Langer particle counter, A. E. particle sizer
Attenuation	Ultrasonic wave particle sizer, x-ray particle sizer
Piezoelectric phenomena	Piezoelectric particle sensor (detects momentum)

[a]Trade name.

diffraction method, for example, provides one of the instruments. However, these instruments have some restrictions in application and are expensive.

7.2 DIRECT CLASSIFICATION METHOD

Particles sampled from powder handling processes are classified into several groups according to their sizes, as in the case of laboratory sieve analysis. On-line instruments detect the powder flow rates or particle concentrations of each size group and calculate the particle size distributions or the mean particle sizes. Therefore, the accuracy depends on performances of classifiers and flowmeters (Chapter 4) or concentration meters (Chapter 5).

The performance of a classifier is represented by a partial separation efficiency which gives the separation efficiency as a function of particle size. Particle size giving 50% separation efficiency is called 50% cut size and is the most important parameter. The separation efficiency around 50% cut size should change sharply with particle size. The sharpness is represented by a quantity such as D_{p75}/D_{p25}, where D_{p75} and D_{p25} are 75% and 25% cut size, respectively. The relation $D_{p75}/D_{p25} = 1$ indicates ideal stepwise separation and, in this case, $D_{p25} = D_{p50} = D_{p75} = D_{pc}$. These sharp cut classifiers associated with flowmeters must be combined in series to obtain the particle size distributions. It is, however, too expensive to introduce such a size sensor in industrial processes. Therefore, it is usual to obtain only an indication of fineness by use of a single-stage classifier [1].

Figure 7.1 shows an example of the measuring system. Sampled powder flow rate W_0 and the flow rate of classified coarser particles W_c are measured by a scale of loss-in-weight mode and a belt scale, respectively. These flow rate signals are fed to a microcomputer to calculate the fraction r_c of coarser particles in the sampled powder.

$$r_c = \frac{W_c}{W_0} \qquad\qquad (7.2.1)$$

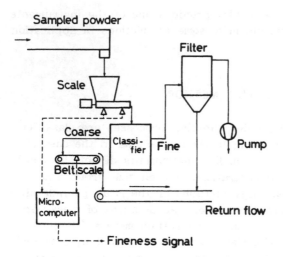

Figure 7.1 Particle sizer based on direct classification.

The value r_c gives only one point on the cumulative size distribution of the sampled powder at the cut size D_{pc}. However, the
value can be utilized as an indication of fineness of powder if
the cut size is adequately selected. It may also be possible to
relate r_c with the following specific surface area diameter D_{sv}
described in Section 1.4:

$$D_{sv} = \frac{6}{S_v} \qquad\qquad (7.2.2)$$

where S_v is the specific surface area (m^2/m^3).

The cut size must be located near the 50% size of the
cumulative size distribution under normal operating conditions in
order to detect the fineness of powder effectively. It is also
possible to change the cut size of classifier so as to maintain
the separation efficiency at 50% by means of feedback control [2];
then the cut size represents the fineness of powder.

Such an instrument for particle size analysis (particle sizer)
has been developed even for micrometer range particles by use

of a centrifugal air classifier [3]. Hopper scales are utilized in
the commercially available sizer, and measurement is carried out
periodically (Nisshin Engineering Co., Japan). Another sizer
of this type utilizes two cyclones in series. Flow rates of powder
collected by cyclones are continuously measured by use of belt
scales (Alpine Co., Germany).

Electrostatic current, described in Section 4.5, is also
utilized to measure powder flow rates (Bradford University
Research Ltd., England). The sizer is for monitoring average
size from 2 to 6 μm in dust-laden gases. A sample of gas-solid
suspension is passed through a cyclonic concentrator and classified into two streams, suspending finer and coarser particles,
respectively. The mass concentrations of powder in the two
outgoing streams are monitored by electrostatic currents. An
ac component is utilized in the sizer because it is strongly
dependent on powder concentration but little affected by electrical
leaks or probe surface conditions.

A single-turn helical channel of rectangular cross section
is utilized to obtain the fineness of particles in a slurry [4].
The concentration of particles near the inner and outer walls of
the channel is measured just downstream of the helix by means
of the beta-ray attenuation technique discussed in Section 5.4.
The classification efficiency depends on the flow in the helix,
and it may be important to suppress the secondary flow, which
causes the sharpness of separation to deteriorate [5].

The particle sizers discussed above detect only one parameter of the size distribution. The following discussion concerns
the method for detecting size distributions. There are two ways
to obtain the size distribution by use of a single-stage classifier.

1. If particle trajectories in the classifier are uniquely determined
    by their respective size, the particle size distribution can be
    estimated by means of local concentrations (the concentration
    scanning method).
2. If the 50% cut size of the classifier is automatically changed
    in a relatively short period, the particle size distribution
    can be determined through continuous measurement of powder
    flow rates (the cut-size sweep method).

The words "local concentrations" in description 1 should be replaced by "local mass fluxes" in a rigorous measurement. It is, however, difficult to obtain the local mass fluxes by noncontact sensing methods. Therefore, the mass fluxes are estimated by use of local concentrations and local particle velocities. The local particle velocities are usually not measured because the measuring system becomes too complicated. The bias of the sensor caused by the local velocity differences is compensated by introducing correlation factors.

Although there is such ambiguity in method 1, the measurement error will be neglected for particles suspended in liquids, because the particle velocity is almost equal to that of fluids. Further, the velocity of fluids can be obtained by numerical calculations based on the Navier-Stokes equation.

For example, if the flow in the helix discussed above is ideal, the method may be extended to obtain the particle size distribution by introducing the slurry only in the neighborhood of the inside wall. Particles in the slurry thus introduced will have different trajectories according to their respective inertia. Therefore, concentration distribution of outgoing stream can be transformed into the corresponding size distribution through some calculations. The concentration distribution may be obtained by beta-ray scanning. The light attenuation technique will also be utilized for larger particles (> 2 μm), as discussed in Section 5.4. The helix may be replaced by some inertia classifiers, although it may be difficult to attain large-particle inertia in liquid flows. The method will also be applied to gas-solid suspension flow as far as the above-mentioned local velocity differences are taken into consideration. A rectangular jet virtual impactor [6] may easily be modified to obtain such a classifier applied for particle sizers. Airflows containing particles are sandwiched by clean airflows in the impactor, and the particle trajectories are uniquely determined by the particle inertia parameter described in Section 4.4.

Figure 7.2 shows an example of method 2. Sampled particles are classified by an air elutriator. Flow rates of particles into the elutriator and of classified coarse particles are continuously measured by impact flowmeters, discussed in Section 4.3. The 50% cut size can be changed by varying the air velocity in the

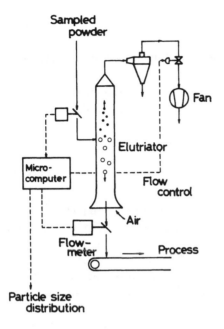

Figure 7.2 Particle sizer based on cut-size sweep method.

elutriator. Therefore, the cumulative oversize distribution can
be obtained in one cycle of sweep by varying the airflow rate [7].
     Classifiers suitable for method 2 are restricted to only a
few types because of the difficulty in changing the cut size.
It is also one of the reasons that the time lag caused by particle
inertia, discussed below, is inevitable during the cut-size sweep.
     The equation of motion for a particle in an elutriator is
given by the following equation based on Stokes' law:

$$m_p \frac{dv}{dt} = 3\pi\mu D_p(u - v) - m_p \underline{g} \qquad (7.2.3)$$

where m_p is mass of the particle, u the air velocity, v the
particle velocity, μ the air viscosity, D_p the particle diameter,
and $\underline{g}$ the gravity acceleration. Dividing both sides of Eq.
(7.2.3) by $3\pi\mu D_p$, the following equation will be obtained.

$$\tau \frac{dv}{dt} = u - v - \tau \underline{g} \tag{7.2.4}$$

where τ is the particle relaxation time introduced in Section 4.4 [Eq. (4.4.5)].

For a given air velocity $u = u_1$, the cut size will be determined by the relation $v = 0$ or $\tau \underline{g} = u_1$. That is, the cut size D_{p1} is given by

$$D_{p1} = \sqrt{\frac{18\mu u_1}{\rho_p \underline{g}}} \tag{7.2.5}$$

If the air velocity u is decreased from u_1 to u_2, the cut size decreases from D_{p1} to D_{p2}. The change in cut size is, however, not immediate, because the particle velocity change is controlled by Eq. (7.2.4), which is a system of first-order time lag. The time required for a particle of size D_{p2} to decelerate to 63.2% of the initial velocity after changing the air velocity stepwise from u_1 to u_2 is given by the particle relaxation time.

The discussion above is based on Stokes' law because of its simplicity. Modifications will be necessary in the fluid drag term if the particle is large.*

---

*The fluid drag acting on a particle is a function of the particle Reynolds number:

$$Re_p = \frac{D_p |u - v| \rho}{\mu}$$

The fluid drag coefficient by Stokes' law is given by the following equation:

$$C_D = \frac{24}{Re_p}$$

The following equation can be utilized for a higher particle Reynolds number:

$$C_D = \frac{24(1 + 0.125 Re_p^{0.72})}{Re_p}$$

In addition to a consideration of the time lag discussed above, it is necessary that the air velocity (or cut size) be smoothly changed. Otherwise, the computer calculations become complicated and the measured size distributions will be severely affected by various noises. Centrifugal air classifiers may be applied successfully to the cut-size sweep method, because the cut size is easily changed by changing rotor speed.

7.3 PERMEABILITY METHOD

The fineness of particles can be represented by means of pressure drop caused by air or liquid flow through a particle bed. The particle bed can be modeled by many small channels through which fluid flows. Those channels are assumed to be constructed by the same length L and the same diameter D_e. The diameter is called the equivalent diameter and defined by the following equation:

$$\frac{D_e}{4} = \frac{\text{cross-sectional area of channels}}{\text{perimeter of channels}}$$

$$= \frac{\text{volume of voids in particle bed}}{\text{total surface area of particles}}$$

$$= \frac{\text{void fraction of particle bed}}{\text{total surface area of particles/volume of particle bed}}$$

$$= \frac{\varepsilon}{S_V} \qquad\qquad (7.3.1a)$$

The total surface area of particles divided by the volume of the particle bed is related to the specific surface area S_V discussed in Section 1.4 as follows:

$$S_V = S_v(1 - \varepsilon) \qquad\qquad (7.3.1b)$$

Therefore,

$$D_e = \frac{4\varepsilon}{(1 - \varepsilon)S_v} \qquad\qquad (7.3.2)$$

The pressure drop can be represented by the Fanning equation:

$$\Delta P = f \frac{2\rho u_c^2 L}{D_e} \tag{7.3.3}$$

where u_c is flow velocity in the channels. The velocity is represented by means of superficial velocity u as follows:

$$u_c = \frac{u}{\varepsilon} \tag{7.3.4}$$

The Fanning friction factor f for laminar flow is given by

$$f = \frac{16}{Re} = \frac{16\mu}{u_c \rho D_e}$$

$$= \frac{4\mu S_v(1 - \varepsilon)}{u\rho} \tag{7.3.5}$$

where Re is the flow Reynolds number.

From Eqs. (7.3.2) to (7.3.5), the pressure drop is represented by the following equation:

$$\Delta P = 2 \frac{(1 - \varepsilon)^2 \mu L u S_v^2}{\varepsilon^3} \tag{7.3.6}$$

In the permeability method, the following Kozeny-Carman equation is utilized instead of Eq. (7.3.6):

$$\Delta P = k \frac{(1 - \varepsilon)^2 \mu L u S_v^2}{\varepsilon^3} \tag{7.3.7}$$

where k is called the Kozeny constant ($\cong 5$) and L is approximated by the height of the particle bed.

From Eq. (7.3.7), the specific surface area is given by the following equation:

$$S_v = \sqrt{\frac{\varepsilon^3}{k\mu L(1 - \varepsilon)^2} \frac{\Delta P}{u}} \tag{7.3.8}$$

If the specific surface area S_v is obtained by means of Eq. (7.3.8), the specific surface area diameter D_{sv} is calculated by Eq. (7.2.2).

There are several ways to obtain the specific surface area. For example:

1. The pressure drop ΔP and the velocity u are measured under the constant-flow condition. The velocity is very slow but it can be measured by use of a capillary (Lee-Nurse's method).

2. A constant volume of air goes through the particle bed under a constant-pressure difference. The specific surface area can be obtained by measuring the time needed for the permeation.

3. A constant volume of air goes through the particle bed as in the case of method 2, but it is not necessary to keep the pressure difference constant. Therefore, the pressure difference and the velocity may change during the permeation. The specific surface area can be obtained by use of the time needed for permeation (Blaine's method).

Blaine's method is based on the fact that Eq. (7.3.8) is valid even for unsteady flow through the particle bed. If the time change of pressure difference is determined by the permeated gas volume V(t), the velocity u in Eq. (7.3.8) is represented by the time derivative of the pressure difference as derived below.

It may be assumed that the flow is laminar and the pressure difference change gives the permeated volume:

$$\Delta P_0 - \Delta P = aV(t) \tag{7.3.9}$$

where a is a proportional constant. On the other hand, the velocity u is given by

$$u = \frac{1}{A}\frac{dV(t)}{dt} \tag{7.3.10}$$

where A is the cross-sectional area of the particle bed. From Eqs. (7.3.9) and (7.3.10), the velocity is represented by use of the pressure difference.

$$u = - \frac{1}{aA} \frac{d(\Delta P)}{dt} \qquad\qquad (7.3.11)$$

Substituting Eq. (7.3.11) into Eq. (7.3.8), the following equation will be obtained:

$$S_v^2 \frac{d(\Delta P)}{\Delta P} = - \frac{aA\varepsilon^3}{k\mu L(1 - \varepsilon)^2} dt \qquad\qquad (7.3.12)$$

Integration of Eq. (7.3.12) from $\Delta P = \Delta P_0$ to ΔP_t and $t = 0$ to t gives the following equation:

$$S_v = \sqrt{\frac{aA\varepsilon^3 t}{k\mu L(1 - \varepsilon)^2 \ln(\Delta P_0/\Delta P_t)}} \qquad\qquad (7.3.13)$$

In Eq. (7.3.13), all parameters except the void fraction ε and time t are determined uniquely through calculations.

If the pressure difference is kept constant during the permeation (method 2), the permeation velocity is given by the following equation:

$$u = \frac{V}{At} \qquad\qquad (7.3.14)$$

Therefore, Eq. (7.3.8) reads

$$S_v = \sqrt{\frac{A\varepsilon^3 \Delta P\, t}{k\mu L(1 - \varepsilon)^2 V}} \qquad\qquad (7.3.15)$$

Method 2 based on Eq. (7.3.15) is also referred to as Blaine's method [8].

Automatic permeability meters utilize Blaine's method (method 2). Powder sampled from a process line is packed into a cell. If the powder is hot, it is cooled in the sampling conveyor before packing. Then the cell moves to the next position and gas permeation starts with the opening of the three-way valve as shown in Fig. 7.3. The permeation gas (usually air) passes through the powder bed as sealing liquid (usually water) goes down. The liquid receiver is opened to the atmospheric pressure.

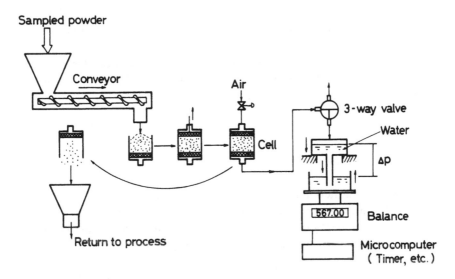

Figure 7.3   Automatic permeability meter.

Therefore, the pressure difference during the permeation is given
by the difference of liquid levels, which is held constant by
lowering the receiver.   The permeated gas volume V is determined
from the weight of the displaced liquid detected by a balance
[8,9].   The time required for the permeation is measured by
use of a timer and the specific surface area can be calculated by
a microcomputer.   The temperature effect on the air viscosity is
also corrected.

The void fraction of the particle bed is usually kept constant
by packing a constant mass of powder into the cell under constant pressure.   It is, however, also possible to utilize the
electric capacitance described in Section 4.5.3 in order to maintain
the constant void fraction.   Powder is packed into the cell under
vibration until the electric capacitance of the particle bed reaches
that of a standard sample [10].   The problem with this method
is that the packing fraction attainable is not enough for fine
particles.   If the packing is too loose, flow through the bed
depends on the aggregated particles instead of the original

particles and the Kozeny constant differs from 5. Also, the
specific surface area obtained does not represent the fineness of
powder.

A pressure response for pulsating flow through the powder
bed is also applied to obtain the void fraction, because the amplitude attenuation of the pulsating pressure depends on the
void fraction [11,12].

7.4 OPTICAL METHOD

The optical method has the feature that it is extremely sensitive,
nearly instantaneous, and does not require physical contact with
the measured particles. A light beam incident on a spherical
particle will be scattered as shown in Fig. 7.4. The typical
intensity distribution of the scattered light is represented in
Fig. 7.4. The scattering angle θ in the scattering plane shown
in the figure is measured from the direction of the incident beam
to the scattered beam. Scattering with $\theta = 0°$ is called forward
scattering, while $\theta = 180°$ is back scattering. I_1 is the component of scattered intensity that is polarized perpendicular to
the scattering plane, and I_2 is for the component parallel to
the plane. The component I_2 will be zero at $\theta = 90°$ and the
light scattered at 90° will be completely polarized perpendicular
to the scattering plane.

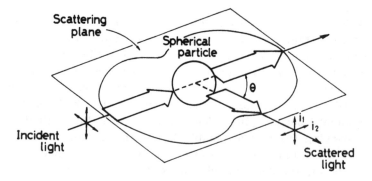

Figure 7.4 Light scattering.

The light scattering is described by Maxwell's theory of electromagnetic radiation, which was solved by G. Mie (1908). The subject discussed here is elastically scattered light, whose frequency (wavelength) is the same as that of the incident light. The intensity I divided by particle volume v_p changes with particle diameter as shown in Fig. 7.5. If the particle size is much smaller than the wavelength λ, I/v_p is proportional to D_p^3 or I is proportional to D_p^6. The scattering is called Rayleigh scattering. For $D_p \cong \lambda/2$, a change in the intensity I is very complicated (Mie resonance). I is proportional to D_p^2 for larger particles, where diffraction of light dominates.

Now we introduce the dimensionless parameter

$$\alpha = \frac{\pi D_p}{\lambda} \qquad\qquad (7.4.1a)$$

which is the ratio of particle circumference to wavelength, called the size parameter. A parameter β is also introduced for convenience:

$$\beta = m\alpha \qquad\qquad (7.4.1b)$$

where m is the refractive index of the particle.

At a distance R in the direction θ from the particle illuminated with unpolarized light of intensity I_0, the scattered intensity $I(\theta)$ is given by the following equation:

$$I(\theta) = \frac{I_0 \lambda^2 (i_1 + i_2)}{8\pi^2 R^2} \qquad\qquad (7.4.2)$$

where i_1 and i_2 are the Mie intensity parameters for scattered light with perpendicular and parallel polarization, respectively. The Mie intensity parameters are given by the following equations:

$$i_1 = \left| \sum_{n=1}^{\infty} \frac{2n + 1}{n(n + 1)} (a_n \pi_n + b_n \tau_n) \right|^2 \qquad\qquad (7.4.3a)$$

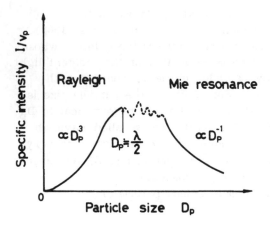

Figure 7.5 Specific intensity of scattered light as a function of particle size.

$$i_2 = \left| \sum_{n=1}^{\infty} \frac{2n + 1}{n(n + 1)} (b_n \pi_n + a_n \tau_n) \right|^2 \tag{7.4.3b}$$

where

$$a_n = \frac{S_n'(\beta) S_n(\alpha) - m S_n'(\alpha) S_n(\beta)}{S_n'(\beta) T_n(\alpha) - m T_n'(\alpha) S_n(\beta)} \tag{7.4.4a}$$

$$b_n = \frac{m S_n'(\beta) S_n(\alpha) - S_n'(\alpha) S_n(\beta)}{m S_n'(\beta) T_n(\alpha) - T_n'(\alpha) S_n(\beta)} \tag{7.4.4b}$$

and

$$S_n(\alpha) = \sqrt{\frac{\pi \alpha}{2}} \ J_{n+1/2}(\alpha) \tag{7.4.5a}$$

$$T_n(\alpha) = S_n(\alpha) + j(-1)^n \sqrt{\frac{\pi \alpha}{2}} J_{-(n+1/2)}(\alpha) \qquad j = \sqrt{-1} \tag{7.4.5b}$$

$$\pi_n = \frac{1}{\sin \theta} P_n^{(1)} (\cos \theta) \qquad\qquad (7.4.6a)$$

$$\tau_n = \frac{\partial}{\partial \theta} P_n^{(1)} (\cos \theta) \qquad\qquad (7.4.6b)$$

J is the Bessel function, $P_n^{(1)}$ is the Legendre polynomial, and S_n' and T_n' are the derivatives of S_n and T_n, respectively.

The extinction efficiency Q_e described in Section 5.4 is represented by

$$Q_e = Q_s + Q_a \qquad\qquad (7.4.7)$$

where Q_s is the scattering efficiency and Q_a is the absorption efficiency. Q_s and Q_e are given by the following equations:

$$Q_s = \frac{\lambda^2}{2\pi} \sum_{n=1}^{\infty} (2n + 1)(|a_n|^2 + |b_n|^2) \qquad\qquad (7.4.8a)$$

$$Q_e = \frac{\lambda^2}{2\pi} \sum_{n=1}^{\infty} (2n + 1)[Re(a_n + b_n)] \qquad\qquad (7.4.8b)$$

where $Re(a_n + b_n)$ denotes the real part of the complex function $(a_n + b_n)$.

Figure 7.6 shows $(i_1 + i_2)$ calculated as a function of the size parameter α [13]. The curves fluctuate violently in the Mie resonance region. Light-scattering instruments usually gather scattered light over a range of scattering angle θ in order to get smooth curves.

The methods based on the light scattering can be subdivided into two groups: those that measure a large number of particles simultaneously, and those that count and size individual particles one at a time. The former type usually measure the angular distribution of light scattered from a large number of particles present simultaneously in a small measurement volume. The angular distribution should be unfolded in order to give the size distribution. The latter type is usually called the particle counter. It requires sampling of aerosol flow. The sample is diluted and constrained to flow through a small jet (=1 mm in

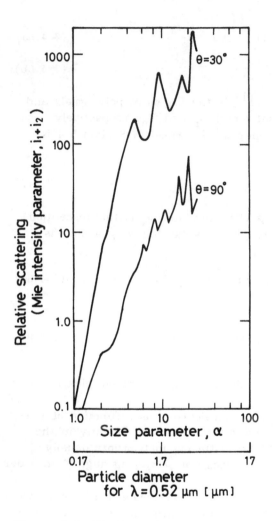

Figure 7.6 Mie intensity parameter as a function of size parameter (water droplets, m = 1.33). (After Ref. 13.)

diameter) and is uniformly illuminated with a collimated light beam.
The scattered light pulses are detected, and pulse heights are
analyzed to give a size distribution on a number basis. The main
problem with such counters is the achievement of a response that
is insensitive to particle refractive index.

   The next discussion is concerned with the diffraction of
light by a suspended particle. When a spherical particle larger
than 1 μm is illuminated as shown in Fig. 7.7 by a parallel laser
beam (monochromatic, coherent light), a diffraction pattern will
be formed. Undiffracted light is focused to a point on the focal
plane and the diffracted light forms a pattern of rings around
the center spot. Movement of the particle does not cause movement of the diffraction pattern, because the light diffracted at
an angle will give the same radial displacement in the focal plane
irrespective of the particle's position in the illuminating beam
[14]. Lens B in Fig. 7.7, called the Fourier transform lens,
brings the far-field diffraction pattern back to the focal plane
[15]. The far-field diffraction pattern thus produced is known
as the Fraunhofer diffraction pattern, for which the intensity
distribution $I(\theta)$ can be described by the following equation:

$$I(\theta) = \frac{\alpha^2}{4\pi}\left[\frac{2J_1(\alpha\theta)}{\alpha\theta}\right]^2 \tag{7.4.9}$$

Figure 7.8 shows $[2J_1(\alpha\theta)/\alpha\theta]^2$ as a function of $\alpha\theta/\pi$. The
diffraction angle θ is approximated by

$$\theta = 1.22\frac{\pi}{\alpha} = 1.22\frac{\lambda}{D_p} \tag{7.4.10}$$

   The light-intensity distribution would appear on a screen as
a series of concentric alternating light and dark rings. The
rings for particles of different sizes would appear at different
positions as recognized by Eq. (7.4.10). When several different-
sized particles are present simultaneously, superimposed light
distribution would be obtained. For a detector consisting of
concentric annular rings, the energy diffracted into a ring is
obtained by integrating Eq. (7.4.9) as follows:

$$E_i = C \sum_{k=1}^{m} n_k D_{pk}^2 \left\{ [J_1^2(\alpha\theta) + J_0^2(\alpha\theta)]_{r=r_i+d} - [J_1^2(\alpha\theta) \right.$$

$$\left. + J_0^2(\alpha\theta)]_{r=r_i} \right\} \tag{7.4.11}$$

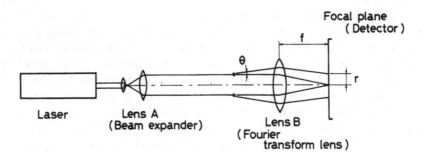

Figure 7.7   Diffraction of laser light by suspended particles.

where r_i and $r_i + d$ refer to the inside and outside radii of the detection ring; C is a constant; J_1 and J_0 are Bessel functions of the first kind of order 1 and 0, respectively; and n_k is the particle size distribution on number basis.

As a similar equation exists for each ring, they can all be summarized by a matrix equation:

$$\underline{E} = \underline{T}\,\underline{n} \tag{7.4.12}$$

where $\underline{E}$ is the light energy distribution (m x 1) vector, $\underline{T}$ is the (m x j) matrix of coefficients that define the light energy distribution curves for each particle, and $\underline{n}$ is the particle size distribution (j x 1) vector on a number basis.

Although Eq. (7.4.12) can be solved for the size distribution vector $\underline{n}$ by use of a computer, considerable computational difficulties will arise because of the large dynamic range of the coefficients, resulting in nonphysical solutions [14]. Therefore, in practical applications, particle size distribution such as Rosin-Rammler distribution is assumed, and the distribution parameters are determined by the method of least squares.

Size measurement by the diffraction method can give erroneous results when the particles are too small (< 1 μm) to satisfy the Fraunhofer diffraction theory. A high concentration of suspended particles (> 10^{11} particles m^{-3}) may cause multiple diffraction, resulting in a smaller size than the actual particle size because of the erroneously large diffraction angle [see also Eq. (7.4.10)].

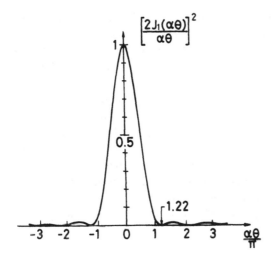

Figure 7.8   Specific intensity distribution for Fraunhofer diffraction.

There is a method that combines the light-scattering method (particle counter) and the light diffraction method. A laser beam is focused by cylindrical lenses to an elliptical cross section. It employs near forward scattering from the focus of the laser beam, together with pulse-height analysis of the signals from individual particles. The method enables measurements with a working space between optical elements of 50 cm. It was anticipated that the technique could be extended to cover particle diameters in the range 0.5 to 50 μm with number concentrations up to 10^{12} m^{-3} [16].

The laser velocimeter described in Section 4.7 is also modified to measure particle size and velocity. Light scattered from two or more coherent beams focused to a common intersection point is observed, with the Doppler difference frequency being used to determine the particle velocity. Relative signal amplitudes are used to determine the particle size. Such devices are referred to as dual scatter laser Doppler velocimeters [17].

The following discussion is concerned with holography, which was developed by D. Gabor in 1948 and extended for

practical applications with laser beams by E. N. Leith & J. Upat-
nieks in 1962 [18].   Figure 7.9 shows the holography system.
A laser beam incidents suspended particles and is partly scattered.
Light scattered by particles interacts with the undiffracted part
of the beam or reference beam, giving rise to interference
fringes which form the hologram on film.   An image of the sample
field can be reconstructed at any future time by illuminating the
hologram with a laser beam, as shown in Fig. 7.9b.   A high-
intensity laser pulse with a duration of 20×10^{-9} s and an output power of about 1 MW was used in the measurement of fine
particles suspended in gas streams [18].   The image reconstructed is three-dimensional, and it is possible to take a
photograph at any plane by use of a camera.   Therefore, particle

a)

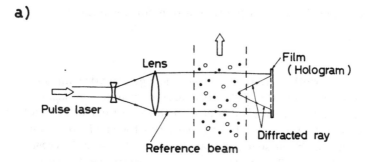

b)

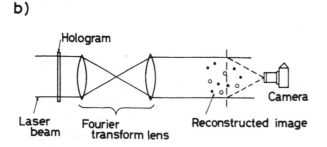

Figure 7.9 Holography system: (a) hologram recording system;
(b) image recording system.

distribution in the sample field as well as the particle size distribution can be obtained by analyzing photographs focused at various positions. Particle velocity can also be estimated by the use of two holograms recorded at a certain interval of time. Further, the particle trajectories will be obtained by analyzing many holograms. There may be, however, difficulties in the image analysis. One technique provides a method of distinguishing each particle easily by the use of two pulse lasers of different wavelength [19]. Two holograms are recorded in two different films separately at a desired time interval, and displacements of moving particles are measured on a superimposed picture of these two films.

7.5 ELECTRICAL METHOD

7.5.1 Induction Charging

Induction charging, discussed in Section 5.3.5, can be applied to particle size measurement. A probe constructed by the electrical system shown in Fig. 5.1a was studied to measure airborne cloud drop size distributions [20]. A pin hole in the upstream tip of the probe faced the metal target rod. The inside volume of the probe was kept under sufficient vacuum to produce airflow at sonic velocity within the pinhole. The insulated target rod was charged to about 400 V or more. There was thus a potential gradient of about 5 kV/cm in the 750-μm gap between probe cap and target. It was possible to count and size cloud droplets in the range 2 to 60 μm at rates up to 10^4 particles per second.

The system shown in Fig. 5.1b was also applied in order to measure drop size distribution [21,22]. Figure 7.10 shows the equivalent electric circuit for the system. C is a condenser with capacitance C representing a particle, C_0 a condenser with capacitance C_0 between the system and the earth, E the applied voltage, R a contact resistance between the particle and the probe, and R_0 a resistance for detecting the current I. If the particle contacts the probe (switch S is closed), the condenser C is charged by Q_2 and C_0 by $Q_1 - Q_2$. Now, we consider the case R = 0. Then the following equations will be obtained:

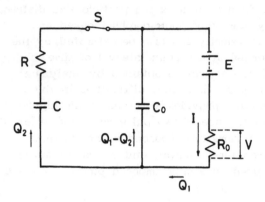

Figure 7.10 Equivalent electric circuit for induction charging.

$$E - IR_0 = \frac{Q_1 - Q_2}{C_0} \qquad (7.5.1)$$

$$\frac{Q_1 - Q_2}{C_0} = \frac{Q_2}{C} \qquad (7.5.2)$$

$$I = \frac{dQ_1}{dt} \qquad (7.5.3)$$

Equations (7.5.1) to (7.5.3) can be solved under the initial condition, $Q_1 = C_0 E$ and $Q_2 = 0$ at $t = 0$, giving

$$I = \frac{CE}{(C_0 + C)R_0} \exp\left[-\frac{t}{(C_0 + C)R_0}\right] \qquad (7.5.4)$$

Therefore, the following voltage will be detected:

$$V = \frac{CE}{C_0 + C} \exp\left[-\frac{t}{(C_0 + C)R_0}\right] \qquad (7.5.5)$$

This is a voltage pulse with a maximum value at $t = 0$. The
maximum value V_0 (pulse height) is given by

$$V_0 = \frac{CE}{C_0 + C}$$

(7.5.6)

and therefore

$$C = \frac{C_0 V_0}{E - V_0} \cong \frac{C_0 V_0}{E}$$

(7.5.7)

For an isolated spherical particle, the capacitance C is given by the following equation:

$$C = 2\pi\varepsilon_0 D_p$$

(7.5.8)

where ε_0 is the dielectric constant of air ($= 8.85 \times 10^{-12}$ F/m). From Eqs. (7.5.7) and (7.5.8), the particle diameter is determined by

$$D_p = \frac{C_0}{2\pi\varepsilon_0 E} V_0$$

(7.5.9)

Equation (7.5.9) shows that the particle diameter is proportional to the voltage pulse height.

If the contact resistance R is not assumed to be zero, the similar derivation gives the following equation:

$$V_0 = \frac{C}{C_0} E f_1(a)$$

(7.5.10)

where

$$f_1(a) = a^{a/1-a} \quad \text{and} \quad a = \frac{CR}{C_0 R_0}$$

(7.5.11a,b)

The parameter a is the ratio of time constants of the charging (CR) and measuring systems ($C_0 R_0$). For electrically conductive particles, $a = 0$. The function $f_1(a)$ is a monotone decreasing function with $f_1(0) = 1$ and $f_1(\infty) = 0$. Therefore, Eq. (7.5.10) means that the contact resistance causes a decrease in the pulse height.

On the other hand, an electrical image effect causes an increase in the pulse height. If the tip of the probe is approximated by a hemisphere of diameter D, the effect is also represented by Eq. (7.5.10), with the following function f_2 instead of f_1:

$$f_2(b) = 1 + (1 - b) \sum_{n=1}^{\infty} \left(\frac{1}{n + b} + \frac{1}{n + 1 - b} - \frac{2}{n} \right) \quad (7.5.12)$$

where

$$b = \frac{D_p}{D_p + D} \quad (7.5.13)$$

Function $f_2(b)$ is a monotone increasing function with $f_2(0) = 0$ and $f_2(1) = 1$. If $D_p \gg D$, the measurement is not affected by the image effect.

7.5.2 Unipolar Charging by Corona Discharge

Particles passing through a corona discharge field are electrically charged by ions existing in the field. The amount of charge on any particle can be determined by the particle size and the conditions of corona discharge. Ions supplied by the corona discharge are accelerated by the external electric field and bombard the suspended particles. This mechanism of charging, called field charging, is best applied to particles larger than about 2 μm. For smaller particles, thermal movement of ions will affect the charging. The second mechanism is called diffusion charging, and particles smaller than 0.2 μm are charged by this mechanism.

Ion current i (A) given by the following equation will be supplied to the surface of a particle suspended in a corona discharge field:

$$i = eNBEA(q) \quad (7.5.14)$$

where N is the concentration of unipolar ions (m^{-3}), e the elementary charge $[1.6 \times 10^{-19} \text{ (C)}]$, B the mobility of ions

$(m^2/V \cdot s)$, E the electric field strength (V/m), and A(q) a correction factor called the effective collision area (m^2).

The effective collision area is given by the following equation:

$$A(q) = \frac{3\varepsilon_s}{\varepsilon_s + 2\varepsilon_0} \pi \frac{D_p^2}{4} \left[1 - \frac{q}{(3\varepsilon_s/\varepsilon_s + 2\varepsilon_0)\pi\varepsilon_0 ED_p^2} \right]^2 \qquad (7.5.15)$$

where q is the charge on the particle (C) and ε_s is the dielectric constant of the particle (F/m).

The charge q changes with the elapse of time, and the ion current will finally cease as the charge equilibrium is attained. The process is represented by the following equation:

$$\frac{dq}{dt} = i = eNBEA(q) \qquad (7.5.16)$$

Equation (7.5.16) can be solved with an initial condition of q = 0 at t = 0:

$$q = \frac{3\varepsilon_s}{\varepsilon_s + 2\varepsilon_0} \pi\varepsilon_0 ED_p^2 \frac{eBNt/4\varepsilon_0}{1 + (eBNt/4\varepsilon_0)} \qquad (7.5.17)$$

The maximum charge q_∞ is given by the following equation as $t \to \infty$ in Eq. (7.5.17):

$$q_\infty = \frac{3\varepsilon_s}{\varepsilon_s + 2\varepsilon_0} \pi\varepsilon_0 ED_p^2 \qquad (7.5.18)$$

As the particle charge q is proportional to D_p^2, the total charge per unit mass of powder (specific charge) q' is proportional to the specific surface area S_w (m^2/kg). Therefore, $6/\rho_p q'$ is proportional to the specific surface area diameter D_{sv}. The total charge of powder can be measured by use of a Faraday cage. Figure 7.11 shows an example of a Faraday cage with a particle filter inside it. The mass of powder collected on the filter can be measured by various methods described in Chapter 5.

Charged aerosol

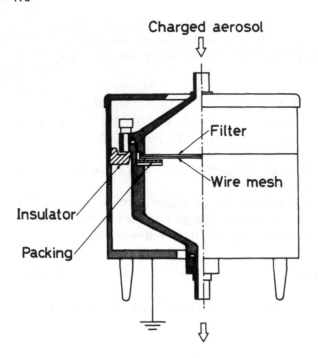

Figure 7.11   Faraday cage with a particle filter.

   Diffusion charging may affect particles smaller than 2 μm, and the charge q of particles smaller than 0.2 μm is nearly proportional to particle diameter instead of D_p^2 [23].   The details are omitted here, however.

7.6   ACOUSTIC METHODS AND OTHERS

7.6.1   Acoustic Method

Acoustic emission caused by particle movement will also be utilized in the measurement of particle size.   It will easily be detected by use of a microphone.   There are two types of acoustic method, one for the measurement of aerosol particles and the other for bulk solids.   Sound attenuation is described in Section 7.6.2.
   An aerosol particle flowing through a capillary causes acoustic emission.   Particles are passed through a capillary 1.5

to 5.0 mm in diameter and 6 cm long in which they are gradually
accelerated to about 100 m/s, and then the particles are suddenly
projected into a wide exit cavity. At this point a pressure
pulse is generated by a particle and gives an audible click [24,
25]. Figure 7.12 shows an example of the acoustical sizing
element [26]. The particles should interact with the boundary
layer in the capillary in order to get pressure pulses. As a
particle reaches the laminar boundary layer, it seems to form a
parcel of turbulence because of the velocity difference. The
parcel then floats downstream to the exit and becomes enlarged
enough to break up the laminar jet into turbulence for a moment.
The boundary layer is probably of near-zero thickness at the
capillary entrance, but increases in thickness with the square
root of the distance from the entrance. Eventually, given enough
length, the layer will reach the radius of the capillary, achieving

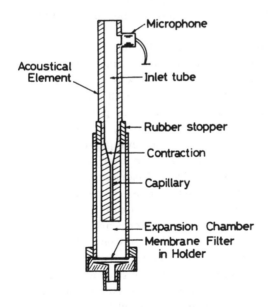

Figure 7.12 Acoustical particle sizing element. [Reprinted with
permission from S. R. Coover and P. C. Reist, Environ. Sci.
Technol. 14, 951 (1980). Copyright 1980 American Chemical
Society.]

the Poiseulle parabolic flow distribution. Somewhere along the
capillary a particle should contact the boundary layer as it
converges into the capillary [25].

The phenomenon can be utilized to count aerosol particles.
The sensor could detect more than 10,000 particles per minute,
but some particles smaller than 30 μm were missed and no
particles smaller than 10 μm were detected by the Langer's
acoustic counter [25]. The detection efficiency E is affected by
flow Reynolds number and particle size [26]. Therefore, if
counts data are gathered at various Reynolds number, the
particle size distribution can be obtained as in the case of Eq.
(7.4.12):

$$\underline{C} = \underline{E}\underline{n} \qquad\qquad\qquad (7.6.1)$$

where $\underline{C}$ is the (m x 1)-vector of counts per unit volume at
various Reynolds numbers Re_1 to Re_m, $\underline{E}$ is the (m x j)-matrix
of detection efficiency for size D_{p1} to D_{pj} at Re_1 to Re_m, and
$\underline{n}$ is the (j x 1)-vector of number of particles per unit volume
at D_{p1} to D_{pj}.

Equation (7.6.1) can be solved if m > j, that is, there are
at least as many counts data at different Reynolds numbers as
size intervals to be determined. The method was applied to
determine the particle size distribution for 6 to 80 μm. Figure
7.13 shows the schematic diagram of the acoustic particle size
analyzer [26]. The acoustic pulse is detected by a miniature
electret microphone and amplified by a linear integrated-circuit
operational amplifier. A voltage comparator is used to discrimi-
nate between acoustical pulses and noise. Discrimination against
noise is readily achieved, as the signal-to-noise ratio is of the
order of 100 : 1. A trigger with presettable dead time, set to
slightly longer than the acoustic pulse duration, prevents multiple
triggering on the sinusoidal waveform of the pulses. A digital
scaler records the acoustic pulses. A membrane filter at the
exit of the particle sizing element prevents contamination of the
subsequent flow system with particulates and dampens mechanical
noise infiltrating back from the vacuum pump [26].

It was also found that the acoustic signal was produced by
a particle entrained in a circular orifice flow instead of the
capillary flow [27]. As larger particles, of order 100 μm, pass

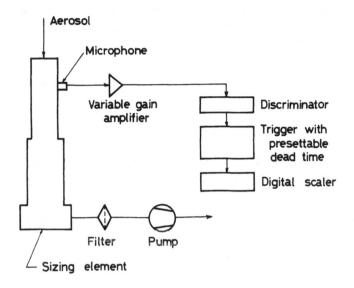

Figure 7.13   Acoustic particle size analyzer.

through the orifice aperture, an audible click is heard.   The
acoustic pressure wave is unique for the shape and size of the
triggering particle.   The pressure wave is derived from the
simple source concept of acoustics theory as follows:

$$p(t) = -\frac{\rho_a}{2\pi r}\frac{dQ}{dt} \qquad\qquad (7.6.2)$$

where $p(t)$ is the pressure wave, ρ_a the air density, r the
observation distance (distance from the orifice to the microphone)
and Q the airflow rate through the orifice.
   The flow rate Q is given by the following equation:

$$Q(t) = U_0[\pi a^2 - B(t)] \qquad\qquad (7.6.3)$$

where B is the blockage area due to the presence of a passing
particle, U_0 the air velocity at the orifice entrance, and a the
radius of the orifice.

The waveform can be calculated by use of Eqs. (7.6.2) and
(7.6.3) when the particle size and shape are assumed. The
following relation is obtained:

$$\frac{rPT}{U_0 \rho_a D_p^2} = \text{const.} \ (= k) \qquad\qquad (7.6.4)$$

where P is the maximum pulse height and 2T is the signal period.

The constant k is about 0.2 to 0.25, and is relatively
independent of particle shape. Therefore, the particle diameter
is estimated by the following equation:

$$D_p = \sqrt{\frac{rPT}{kU_0 \rho_a}} \qquad\qquad (7.6.5)$$

Equation (7.6.5) was found to predict diameters of spherical
particles accurately in the range 25 to 1500 μm [27].

The acoustic method for the measurement of bulk solids is
based on sound originating from rigid particles when they are
interacting elastically. Particles impinging on each other emit
sound signals whose features are related to the size of the
particles. The acoustic frequencies of interest lie in the ultrasonic range. When two isotropic and rigid spheres of the same
diameter and of the same material collide elastically, vibrations
are excited in both spheres, which results in the emission of
acoustic pulses. It is possible to calculate the diameter of
spheres from measurements of the frequencies of vibration.

When the two spheres are slightly different in diameter,
the frequencies of the acoustic signals are related to the arithmetic mean of the fundamental frequencies of the individual
spheres and their harmonics. The collision of such spheres will
result in the generation of beat frequencies which can be orders
of magnitude less than the fundamental frequencies of the
individual spheres.

Figure 7.14 shows a typical acoustic signal produced by a
rotating form-lined cylindrical vessel containing quasi-monodisperse
steel spheres [28]. Two distinct beat patterns are found, one
corresponding to higher-order harmonics and the other to beats
produced by lower-order harmonics. For the higher-order

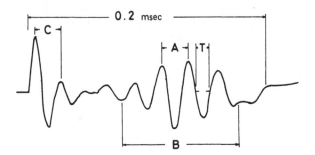

Figure 7.14 Acoustic signal produced by a rotating vessel containing steel spheres. (From Ref. 28.)

hermonics, B is the beat duration, T the half-wavelength within the envelope, and A the wavelength of a steep maximum within the envelope. C is the average wavelength of a steep maximum for the lower-order harmonics.

It was found that the half-wavelength $T(\mu s)$ is directly proportional to the diameter of steel spheres:

$$T = kD_p \qquad (7.6.6)$$

Therefore, the beat frequency is inversely proportional to the diameter. The proportional constant k in Eq. (7.6.6) is the same for other materials as long as the ratio of Young's modulus to density is the same. The wavelengths A and C are also proportional to the diameter D_p. The proportional constants were almost the same for steel spheres and glass spheres.

Further, if these parameters are classified into different mode groups by use of different recorder sweep speeds, the ratio T/B in any fixed mode increases with the width of the size distribution and may be an index of the size distribution.

The acoustic method for bulk solids discussed above is utilized in the control of closed-circuit grinding systems [29]. A microphone is set just outside the mill. If the mill is composed of two compartments, two microphones are utilized, one for each compartment. The acoustic signals from these microphones are analyzed to obtain the average particle size and the mass of holdup powder. It was reported that the control system worked well and the grinding system was stabilized.

7.6.2 Attenuation Technique

As discussed in Section 5.4, energy of waves such as light, sound, or radioactive rays will be reduced by suspended particles because of the absorption or scattering.

Sound Attenuation

Sound attenuation is utilized in the particle size measurement of suspensions. Lambert-Beer's law is also applied to the attenuation of sound pressure:

$$I = I_0 \ \exp \ (-\sigma L) \tag{7.6.7}$$

where I_0 is the intensity of the incident wave, σ the extinction coefficient (also called absorption or attenuation coefficient), and L the distance traveled by the wave.

    Attenuation of sound energy through unit volume of suspension is represented by the following equation [30]:

$$E(t) = E_0 \ \exp \ (-2\sigma a t) \tag{7.6.8}$$

where a is the propagation velocity of sound and t the time, and therefore $a \cdot t$ gives the distance traveled by the sound. The initial energy of sound E_0 is given by the kinetic energy of fluid oscillation:

$$E_0 = \frac{1}{2} \ \rho u_0^2 \tag{7.6.9}$$

where u_0 is the velocity amplitude of fluid oscillation and ρ is the fluid density.

    If the energy loss is caused primarily by the viscous losses associated with the relative movement of fluid and particles, it will be estimated by the following equation:

$$\overline{\frac{dE}{dt}} = -3\pi\mu D_p n \overline{(u - v)^2} \qquad (7.6.10)$$

where $\overline{dE/dt}$ is the rate of energy dissipation per unit volume
of suspension, n the number concentration of particles, u the
fluid velocity, and v the particle velocity. Equation (7.6.10) is
the time-averaged equation.

From Eq. (7.6.8), the following equation will be obtained:

$$\overline{\frac{dE}{dt}} = -2\sigma a \overline{E_0} \exp(-2\sigma at) = -2\sigma a \overline{E} \qquad (7.6.11)$$

If the extinction coefficient σ is constant throughout the propagation, it will be determined by Eqs. (7.6.8) to (7.6.11) as follows:

$$\sigma = -\left.\frac{\overline{dE/dt}}{2a\overline{E}}\right|_{t=0} = \frac{3\pi\mu D_p n \overline{(u - v)^2}}{\rho a u_0^2} \qquad (7.6.12)$$

On the other hand, the following relationship will be obtained
from the equation of motion of a particle:

$$\frac{\overline{(u - v)^2}}{u_0^2} = \frac{1}{2}\frac{\omega^2\tau^2}{1 + \omega^2\tau^2} \qquad (7.6.13)$$

where τ $(= \rho_p D_p^2/18\mu)$ is the particle relaxation time introduced
in Section 4.4 and ω is the angular frequency of the wave. From
Eqs. (7.6.12) and (7.6.13), the extinction coefficient σ is represented by the following equation:

$$\sigma = \frac{3\pi\mu n D_p}{2a\rho}\frac{\omega^2\tau^2}{1 + \omega^2\tau^2} \qquad (7.6.14)$$

If the mass concentration of the suspension $c = (1/6)\pi D_p^3 n \rho_p/\rho$ is
introduced, the attenuation coefficient per wavelength $\bar{\sigma} = \sigma a/\omega$
(unitless) is given by the following equation:

$$\bar{\sigma} = \frac{1}{2}\,c\,\frac{\omega\tau}{1 + \omega^2\tau^2} \qquad\qquad (7.6.15)$$

The maximum attenuation per wavelength will be obtained when $\omega\tau = 1$. Therefore, the frequency for maximum attenuation per wavelength f_0 is given by

$$f_0 = \frac{9\,\mu}{\pi\rho_p D_p^2} \qquad\qquad (7.6.16)$$

A suspension of 10 µm particles in air would have a frequency f_0 of about 515 Hz. Equation (7.6.16) will be rewritten as follows:

$$D_p = 3\sqrt{\frac{\mu}{\pi\rho_p f_0}} \qquad\qquad (7.6.17)$$

On the other hand, if the particle size is held constant, the mass concentration c will be estimated by use of Eq. (7.6.15).

Ultrasonic waves are utilized to measure both the velocity of slurry and the mass median diameter of particles suspended in the slurry [31]. The Doppler shift of the frequency of the wave gives the velocity as in the case of the laser Doppler velocimeter described in Section 4.7. On the other hand, fluctuation of the amplitude of the wave depends on both the particle concentration and the particle size. If the particle concentration is held constant, the cube root of the amplitude fluctuation is directly proportional to the mass median diameter. Therefore, the particle size can be obtained through the intensity measurement of the ultrasonic wave [31].

Further theoretical study will be found in the literature [32] concerning the influence of particle size distributions on the velocity and attenuation of sound in gas-solid suspensions. The influence of a particle size distribution on the sound attenuation is more complicated than its influence on the sound velocity.

X-Ray Attenuation or Fluorescence

The intensity of fluorescent x-rays from a chemical element in a particle depends on particle size. Next we consider a sample

consisting of a binary system of monodispersed cubic particles
suspended in a homogeneous matrix. If the incident x-rays
strike the sample surface at an angle of 45° and the flourescent
x-rays are measured at both an angle of 45° to the surface and
at right angles to the incident beam, the total intensity I of the
fluorescent x-rays from all particles in the sample is given by
the following equation [33]:

$$I(D_p) = \frac{cEJ}{\sqrt{2}B} \frac{\exp(-CD_p)[\exp(AD_p) - 1][\exp(DD_p) - 1]}{ADD_p^2}$$

$$\times \frac{H}{H - \sqrt{2}D_p} [\exp(-BD_p) - \exp(-\sqrt{2}BH + BD_p)]$$

$$(7.6.18)$$

where

$$E = \text{excitation constant}$$
$$J = \text{irradiance of the incidence beam}$$
$$c = \text{holdup volume concentration}$$
$$H = \text{thickness of the sample}$$
$$A, B, C, D = \text{constants determined by the absorption}$$
$$\text{coefficients for different paths of x-rays}$$

The absorption coefficients are dependent only on the
compositions of chemical elements and densities of the two phases
and the wavelengths of the incident and fluorescent x-rays.

Equation (7.6.18) is indeterminate when $D_p = 0$, but
application of l'Hospital's rule gives

$$I(0) = \frac{cEJ}{\sqrt{2}B} [1 - \exp(-\sqrt{2}BH)] \qquad (7.6.19)$$

Since the sample thickness is considerably larger than the particle
size ($H \gg D_p$), the intensity ratio R is given by the following
equation:

$$R = \frac{I(D_p)}{I(0)} \cong \frac{[1 - \exp(-AD_p)][1 - \exp(-DD_p)]}{ADD_p^2} \qquad (7.6.20)$$

Therefore, the mean particle size can be obtained by use of Eq. (7.6.20) if the intensity ratio R is measured.

The intensity ratio R depends also on the particle concentration. Experimental data showed that R decreased significantly when the particle concentration was less than 1 wt %. Above about 1 wt %, however, the intensity ratio was relatively insensitive to particle concentration [33].

Figure 7.15 shows an apparatus being used to measure the sizes of two different types of particles in a process stream. If the concentration of a particular element in a process stream can be controlled or measured, it should be possible to measure the stream through the flow cell as shown in the figure. In

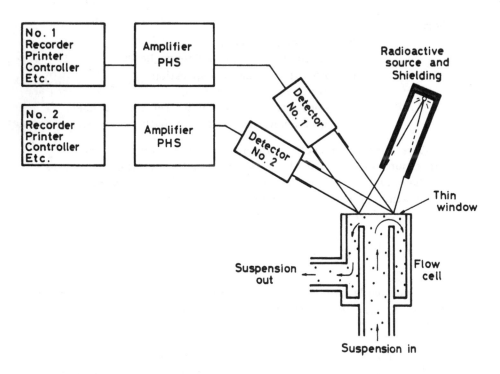

Figure 7.15 On-stream size analysis based on x-ray attenuation. (From Ref. 33.)

cases where the x-ray lines from the chemical elements in the
stream are far apart, the x-ray tube and monochromator could
be replaced by a radioactive source and a pulse-height selector.

7.6.3 Correlation Technique

The correlation technique discussed in Section 2.2 offered one of
the methods to measure the particle velocity and the concentration, as shown in Section 4.6. The technique will also be
applied in the measurement of particle size.

   Particles suspended in turbulent fluid try to follow the
turbulent motion of the fluid. Due to the inertia of the particles,
however, the large particles tend to follow only large eddies,
while small particles can follow both large and small eddies. It
is therefore possible to relate the particle size to a scale of
particle turbulence if the average flow velocity is constant.
The term "particle turbulence" means turbulence observed through
the particle motion. Any signal resulting from the suspension
flow may be utilized to determine the scale of particle turbulence.
For the flow of slurry, the variation in conductivity as particles
passed between two electrodes gives one of those signals [34].
In this method, a multivibrator generates a square wave of 20
kHz, and the variations in electrode conductivity modulate the
amplitude of the 20-kHz carrier wave. This modulated signal
is then rectified so that only the positive portion is used. The
frequency of the carrier wave is stripped off by both a low-pass
filter which has a cutoff value of 1 kHz and a high-pass filter
at 2 Hz in order to remove the dc offset and dc drift. The signal
can then be amplified and the root-mean-square voltage is obtained
by use of a real-time digital correlator. The correlation coefficient $R_E(\tau)$ between the velocity of a particle at time t and
t − τ is given by the following equation:

$$R_E(\tau) = \frac{\overline{u(t)u(t - \tau)}}{\overline{[u(t)]}^2} \qquad (7.6.21)$$

The scale of turbulence can then be obtained through the following integration:

$$T_E = \int_0^\infty R_E(\tau)\ d\tau \cong \int_0^{t_0} R_E(\tau)\ d\tau \qquad (7.6.22)$$

where t_0 is the time delay elapsed until the correlation coefficient $R_E(\tau)$ first becomes zero. The scale of turbulence (the so-called "Eulerian time scale") thus obtained was well represented by the following equation with experimental constants A and B [34]:

$$T_E = A + B\ \ln\overline{D}_p \qquad (7.6.23)$$

where $\overline{D}_p$ is the mean particle diameter (mass median diameter). Therefore, the mean size of particles suspended in the turbulent flow can be estimated by measuring the Eulerian time scale.

Optical signal can also be utilized. Here the simultaneous measurements of velocity, concentration, and the particle size will be introduced. Two 5-mW He-Ne laser beams were projected normal to a slurry flow [35]. The distance between the two laser beams set along the flow was either 2 or 5 mm. Laser light scattered by particles flowing in a pipe was collected at right angles to the beams and focused via two individual lens systems onto a single hybrid photodetector. The random electrical output, m(t), from the hybrid photodetector is the composite signal comprising the scattered light from two laser beams. That is, the signal m(t) is given by the following equation:

$$m(t) = x(t) + y(t) \qquad (7.6.24)$$

where $x(t)$ and $y(t)$ are the outputs from the first and second positions along the pipeline, respectively.

The autocorrelation $\phi_{mm}(\tau)$ is represented by the following equation:

$$\phi_{mm}(\tau) = \phi_{xx}(\tau) + \phi_{xy}(\tau) + \phi_{yx}(\tau) + \phi_{yy}(\tau) \qquad (7.6.25)$$

The autocorrelogram of the output signal m(t) from the hybrid detector has two peaks, as shown in Fig. 7.16. The first peak at $\tau = 0$ is caused by the autocorrelations ϕ_{xx} and ϕ_{yy}, and the second one by the cross-correlations ϕ_{xy} and ϕ_{yx}.

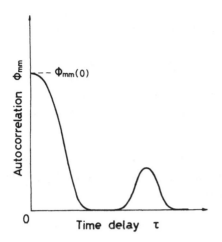

Figure 7.16   Autocorrelogram of m(t) (m = x + y).

      Combined cross- and autocorrelation of the composite electrical signals from the hybrid photodetector were obtained from a correlator for various particle concentrations. The calculated cross-correlation flow velocity was found to be in good agreement with the independent mass method of velocity measurement, irrespective of concentration and particle size.

      On the other hand, the range of frequencies in a turbulent flowing slurry is greater when the slurry contains small particles than when large particles are present. The power spectrum by Fourier transformation of ϕ_{mm} can be used to estimate the mean particle size of the flowing particles from the frequency bandwidth. High-frequency lines of the power spectrum indicate the presence of small particles following small eddies, whereas low-frequency lines represent small and large particles following large eddies. From each individual spectrum an arbitrary value of the power spectra, for example, 50% of the power, can be chosen to give a unique bandwidth for each size distribution. The smaller the mean particle size is, the larger the bandwidth. The bandwidth determined by the Wiener-Khinchine algorithm is not, however, solely dependent on the mean particle size, but is also dependent on the flow velocity. Experimental work showed that the bandwidth was proportional to the flow velocity.

The concentration of slurry is, as discussed in Section 4.6, estimated from the magnitude of the autocorrelogram, $\phi_{mm}(0)$. The value of $\phi_{mm}(0)$ is independent of velocity but dependent on the mean particle size. Therefore, the concentration determined by $\phi_{mm}(0)$ should be corrected by use of the mean particle size.

An optical fiber probe can also be utilized to get optical signals. The size of solid particles moving in particulate equipment can be measured simultaneously with the velocity of particles by use of a small optical fiber probe [36]. An optical fiber probe consists of two optical fibers; one works as a light projector and the other as a receiver. Two probes are set in series with distance L. The average velocity v of the moving particles can be determined through cross-correlation between two received signals. On the other hand, the mean particle size D_p can be approximated by the following equation:

$$\frac{D_p}{D} = 3.2 \left(\frac{s}{D} - 1 \right)^{1.08} \qquad (7.6.26)$$

where D is the fiber diameter and s is the product vt_s of the velocity v and the width of the detected pulse t_s. The method above was tested by use of a rotating cylindrical bed of particles such as glass beads, alumina, or sand [35]. The experimental results showed that the numerical constants in Eq. (7.6.26) depended on the type of particle.

7.6.4 Other Methods

The piezoelectric sensor described in Section 4.5 can be applied in the measurement of the mean particle size. The sensor detects the momentum of moving particles. Therefore, if the particle velocity is held constant, the particle mass will be estimated by analyzing the electric pulse. It was reported that the piezoelectric sensor could be applied in the particle size measurement of powder on a constant-speed belt sampler through detecting the amplitude and frequency of the vibration of the sensor [37].

The electric current generated by electrification of particles discussed in Section 4.5 will also be applied in the particle size measurement of dilute gas-solid suspensions as long as both the velocity and concentration are held constant. The current per unit powder flow rate is inversely proportional to the mean particle size (Sauter diameter).

REFERENCES

1. A. Supp and H. Auerbach, Zem. Kalk Gips 3, 134 (1972).

2. Y. Kawamura, T. Aso, and T. Numata, J. Res. Assoc. Powder Technol. Jpn. 14, 195 (1976).

3. Y. Yamada, M. Yasuguchi, E. Honma, M. Kabaya, H. Tomiyasu, and K. Iinoya, J. Soc. Powder Technol. Jpn. 20, 128 (1983); 21, 627 (1984).

4. B. F. Osborne, ISA Trans. 10, 379 (1971).

5. A. B. Holland-Batt and M. W. Birch, Powder Technol. 10, 189 (1974).

6. H. Masuda, A. Yasuki, and S. Kawaguchi, Kagaku Kogaku Ronbunshu 5, 561 (1984).

7. Y. Nakajima, T. Tanaka, and K. Goto, Kagaku Kogaku 31, 1120 (1967).

8. J. P. Blanc, Zem. Kalk Gips 8, 344 (1968).

9. M. Hasler, Zem. Kalk Gips 3, 132 (1972).

10. E. S. Porter and J. Warshawsky, IEEE Trans. Ind. Gen. Appl. 5, 773 (1969).

11. Y. Nakajima, Y. Ado, K. Gotoh, and T. Tanaka, Ind. Chem. Fundam. 9, 489 (1970).

12. Y. Nakajima and T. Tanaka, Ind. Chem. Eng. Fundam. 10, 318 (1971).

13. W. C. Hinds, Aerosol Technology, John Wiley & Sons, Inc., New York, Chap. 16 (1982).

14. B. J. Azzopardi, Filtr. Sep. 21, 415 (1984).

15. C. Yamanaka (ed.), Laser Handbook (Japanese), Ohmsha, Ltd., Tokyo, p. 109 (1982).

16. D. Holve and S. A. Self, Appl. Opt. 18, 1632 (1979).

17. W. M. Farmer, Appl. Opt. 15, 1984 (1976).

18. A. W. Jury, E. J. Addis, and G. P. Reed, Powder Technol. 38, 39 (1984).

19. T. Murakami, M. Ishikawa, M. Shirouzu, and T. Shirokane, Zai-Ryou (J. Soc. Mater. Sci. Jpn.) 27, 681 (1978).

20. D. P. Keily and S. G. Millen, J. Meterol. 17, 349 (1960).

21. K. Tamano, Kagaku Kogaku Ronbunshu 7, 390 (1981).

22. K. Tamano, Kagaku Kogaku Ronbunshu 8, 470 (1982).

23. N. A. Fuchs, Geofis. Pura Appl. 56, 185 (1963).

24. G. Langer, J. Colloid Sci. 20, 602 (1965).

25. G. Langer, Powder Technol. 2, 307 (1968); 6, 5 (1972).

26. S. R. Coover and P. C. Reist, Environ. Sci. Technol. 14, 951 (1980).

27. G. M. Bragg and B. C. Morrow, J. Phys. E. 14, 26 (1981).

28. M. F. Leach, G. A. Rubin, and J. C. Williams, Powder Technol. 16, 153 (1977); 19, 189 (1978).

29. K. Iinoya, (ed.), Funryuutai Keisoku Handbook (Handbook for the Instrumentation of Powder Handling Processes), Nikkan Kogyo, Tokyo, p. 260 (1981).

30. S. Temkin, Elements of Acoustics, John Wiley & Sons, Inc., New York, p. 455 (1981).

31. B. Pfau, Verfahrenstechnik (Mainn) 8, 258 (1974).

32. W. Gregor, J. Raasch, and H. Rumpf, Chem. Eng. Sci. 31, 15 (1976).

33. W. A. Hockings, Powder Technol. 3, 29 (1969).

34. M. S. Beck, K. T. Lee, and N. G. Stanley-Wood, Powder Technol. 8, 85 (1973).

35. N. G. Stanley-Wood, G. J. Llewellyn, and A. Taylor, Powder Technol. 29, 217 (1981).

36. K. Oki, T. Akehata, and T. Shirai, Powder Technol. 11, 51 (1975).

37. O. Heinemann, Zem. Kalk Gips 3, 137 (1972).

8

Moisture Content

Methods for the measurement of moisture (water) content of
powders are discussed here. The water held in powder products
is usually included in their price. That is, the water has the
same price as the powder. Therefore, the measurement of
water content is very important even in commercial transactions.
In Section 8.1 we outline the measurement principles and define
water content.

8.1 INTRODUCTION

Powder properties such as the angle of friction or the flowability
of powder depend on the moisture content of the powder. The
moisture content of some powders should be controlled so as to
match storage conditions. If the moisture content is too high,
the powder will deteriorate during storage. On the other hand,

overdrying is a waste of energy. In pressing or plastic treatment of powders, the moisture content affects mold filling,
compressibility, binding properties, and mold ejection. Incorrect
moisture content causes the products to deteriorate. In powder
handling processes, drying is one of the most energy-consuming
processes, and optimization of the system should be carried out
to save energy. The optimization or automatic control of these
processes requires continuous monitoring of moisture content in
the powder materials.

Table 8.1 shows various principles for the measurement of
moisture content, including electrical, optical, nuclear related,
or physicochemical methods. Some of them, especially the
electrical methods, have close relations with Section 4.5.

The moisture content in powder materials is represented by
either w_w on a wet base or w_d on a dry base, as follows:

$$w_w = \frac{M_w}{M} \tag{8.1.1}$$

Table 8.1 Moisture Content Meter

Principle	Method
Electric capacitance increase	Capacitance meter (Sec. 4.5.3)
Electric resistance decrease	Conductivity meter
Microwave attenuation	Microwave moisture meter (Sec. 4.5.5)
Infrared reflection (absorption)	Three color infrared moisture meter
Nuclear magnetic resonance	NMR moisture meter
Neutron reflection (absorption)	High-energy neutron absorption
Cobalt chloride reaction	Color change by reaction
Equilibrium relative humidity	Temperature difference measurement (Chap. 11)

$$w_d = \frac{M_w}{M_d}.$$ (8.1.2)

where M is the mass of the wet powder, M_w the mass of water
contained in the wet powder, and M_d the mass of the fully dried
powder. As the mass of the wet powder M is equal to $M_d + M_w$,
there is the following relationship between the wet base moisture
content and the dry base content:

$$w_d = \frac{w_w}{1 - w_w} \quad \text{or} \quad w_w = \frac{w_d}{1 + w_d}$$ (8.1.3a,b)

When a wet powder is brought into contact with air of lower
humidity than that corresponding to the moisture content of the
powder, the powder tends to lose moisture and dry to the
equilibrium state. The final moisture content is called the
equilibrium moisture content. When the air is more humid than
the powder in equilibrium with it, the powder absorbs moisture
from the air until an equilibrium state is attained. The fraction
of the water in the wet powder that cannot be removed by the
air is called the equilibrium moisture. The difference between
the total water content and the equilibrium water content is
called free water. The water up to the lowest concentration
that is in equilibrium with the air of 100% relative humidity is
called bound water because it exerts a vapor pressure less than
that of liquid water at the same temperature. The water in fine
capillaries (less than 100 Å) exerts lower vapor pressure than
that of liquid water because of the concave curvature of the
surface.* The water bound on the inside surface of pores and
capillaries has properties resembling those of the dry powder
rather than those of free water.

---

*Equation (5.5.7) can also be applied to estimate the vapor
pressure by replacing the particle diameter with a negative value
of capillary diameter and the surface tension γ in the equation
with $\gamma \cos \theta$, where θ is the contact angle between water and
the material.

The water in natural organic substances is in physical and chemical combination. The nature and strength of the combination depends on the moisture content. On the other hand, unbound water exerts its full vapor pressure. The distinction between bound and unbound water depends on the material itself. It is unbound water, rather than bound water, that is of interest in moisture content measurements.

The equilibrium moisture content depends on the temperature, relative humidity, and the material itself [1,2]. It can be approximately represented as a function of $RT \ln(p_s/p)$ for each material, where R is the gas constant (8.314 J/mol·K), T (K) the temperature, p the vapor pressure, p_s the saturated vapor pressure, and p/p_s the relative humidity. The particle size, shape, surface roughness, and packing fraction of particles will also affect the equilibrium moisture content. It is usual, in the complicated situation of water in powder materials that moisture content meters are calibrated by a drying method or Karl Fischer titration.

8.2 MICROWAVE ATTENUATION

The energy of electromagnetic waves will be partly absorbed by water in powders. Absorption at audio and radio frequencies is the result of induced currents in the sample and depends strongly on the conductivity of the sample. At a microwave frequency, absorption is caused by a resonance phenomenon involving rotation of the water molecule. Resonance absorptions also occur at infrared frequencies due to vibrations between the hydrogen and oxygen atoms. Absorption in the optical range is due to vibrations between electrons and atomic nuclei.

The dielectric properties of a wet material in relation with the electromagnetic wave is expressed by use of the complex dielectric constant (relative permittivity). In this chapter, all the dielectric constants are divided by that of air. That is, they are specific dielectric constants. For simplicity, however, the specific dielectric constants are called just dielectric constants in the following sections. Then the complex dielectric constant is represented by the following equation:

$$\bar{\epsilon} = \bar{\epsilon}' - j\bar{\epsilon}'' \qquad j^2 = -1 \tag{8.2.1}$$

where $\bar{\epsilon}'$ is the dielectric constant of the material and $\bar{\epsilon}''$ is called the loss factor.

High loss factors occur in water at a microwave frequency and local absorption appears to peak in the microwave K-band, about 22 GHz, as shown in Fig. 8.1 [3,4]. As the loss factor of water is much higher (more than 100 times) than that of other materials, microwave water measurement may be nearly independent of the influence of base materials. This fact leads to the use of microwaves for moisture measurements.

The ratio of output to incident power of the microwave is given by the following equation as in the case of Section 5.4:

$$\frac{I}{I_0} = \exp(-2\alpha L) \tag{8.2.2}$$

where I_0 is the incident power, I the output power, L the sample thickness, and α the attenuation constant. The attenuation

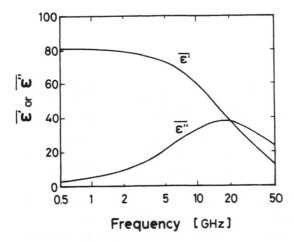

Figure 8.1 Dielectric properties of water as a function of wave frequency in the microwave range (dielectric constant $\bar{\epsilon}'$ and loss factor $\bar{\epsilon}''$).

constant (nonmagnetic dielectrics) is approximated by the following
equation [3,4]:

$$\alpha \cong \frac{\pi \bar{\varepsilon}''}{\lambda \sqrt{\bar{\varepsilon}'}}$$
(8.2.3)

where λ is the wavelength of the microwave. Substituting Eq.
(8.2.3) into Eq. (8.2.2), the attenuation in decibels is obtained
as follows:

$$A = -10 \log \frac{I}{I_0}$$

$$= \frac{27.3 \bar{\varepsilon}''}{\lambda \sqrt{\bar{\varepsilon}'}} L \qquad dB$$
(8.2.4)

Despite the complexity of the physical structure of wet
materials, there usually exists a simple linear relationship
between the moisture content and the attenuation A, at least
in a limited range of moisture content. Losses in dry air are
sufficiently small for the range 1 to 30 GHz, and are ignored in
most cases. The attenuation of a water-material mixture can be
approximated as a sum of individual attenuations:

$$\frac{A}{L} = k_1 \frac{M_w}{V} + k_2 \frac{M_d}{V}$$
(8.2.5)

where V is the sample volume and k_1 and k_2 are constants
calculated by use of Eq. (8.2.4) for water and dry material,
respectively. Equation (8.2.5) shows that the second term can
be neglected in most cases, because the attenuation caused by
the dry material is much less than that of water. The attenuation due to bound water is assumed to be similar to that for
ice, and it is so small that it can be neglected in most cases.
Changes in water from bound to unbound, or vice versa, could
affect measured attenuation. The relaxation frequency of
bound water is much lower than that of liquid water (relaxation
frequency of liquid water is about 17 GHz).

The attenuation also depends on temperature, because the
orientation of water molecules is affected considerably by thermal

agitation. Although the temperature effect depends on the wave
frequency applied, the loss factor in the practical frequency
range decreases with increasing temperature. Samples of fire
clay, sand, and so on, give negative temperature coefficients,
as expected. However, low-moisture-content samples of wheat,
soap, tobacco, chemical fertilizer, and so on, have positive
temperature coefficients (increasing attenuation with increasing
temperature). This fact may be caused by changes in hydrogen
bonding. Increasing temperature increases molecular activity
and bound water changes to unbound. As a result, the ability
to attenuate microwave energy becomes higher. Thus organic
materials have positive attenuation coefficients, whereas non-
hydrogen bonding materials have negative temperature coefficient,
as in the case of liquid water.

Figure 8.2 shows the schematic diagram of a microwave
moisture content meter. The system is composed of a stream-
forming unit for the material, microwave sensors (electromagnetic
horns), microwave transducers, and signal processing units.
Measurement is usually carried out with a constant-frequency
wave. The material stream-forming unit provides proper measuring conditions in the sensor area by attaining the specified

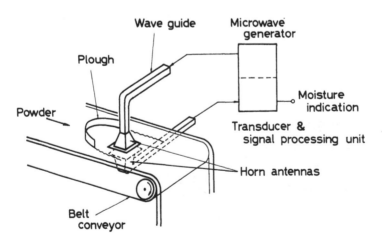

Figure 8.2 Microwave moisture content meter.

powder layer thickness of constant density and flat material surface. Mechanical means such as vibrators, rollers, scrapers, plows, and deflecting plates are utilized for this purpose. The material temperature is also monitored and compensated.

The microwave generated by a source oscillator is transmitted through a wave guide to a horn antenna and then radiated into the powder bed. The microwave penetrated through the powder bed is received by a receiving horn antenna and transmitted through a wave guide to a detector. The output of the detector is controlled by a modulator and the attenuation is determined.

The wave guide is a metal tube with a circular or rectangular cross section. Such wave guides have high transmission efficiency for high-frequency electromagnetic waves such as the microwave. The wave mode TE_{11} is usually selected for circular wave guides and TE_{10} for rectangular wave guides (see also Section 4.5.5). Figure 8.3 shows a rectangular wave guide connected to an electromagnetic horn antenna. The shape of the horn is determined so as to get good directivity and sufficient power gain from the microwave. If a horn of the same shape is used as a receiving antenna, its directivity and power gain are equal to that of the radiating antenna [5].

Inexpensive solid-state devices and integrated circuits have become available for the generation, modulation, switching, and detection of microwave signals, and the cost of installation of microwave moisture content meters has become less expensive. Moisture content calculations are carried out by a microprocessor

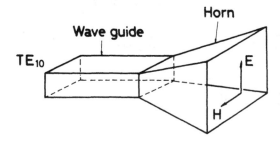

Figure 8.3 Electromagnetic horn antenna.

which also controls the functioning of transducers and other
process operations. Small fluctuations in moisture content are
integrated over time to provide an average moisture content,
which is sometimes more important than the instantaneous value.

Some of the advantages of microwave measuring methods are
as follows [4]:

1. Physical contact between the equipment and the test medium
   is not required.
2. They do not deteriorate nor contaminate the test materials.
3. Microwave radiation is relatively insensitive to the environ-
   mental conditions; thus dust and water vapor do not affect
   measurement accuracy, contrary to infrared methods.
4. As microwaves can propagate through powder samples, the
   volume moisture content can be measured, contrary to
   infrared methods.

The method discussed above utilizes only the value of
attenuation. The primary limitation of such a one-parameter
measurement is the necessity of maintaining the material density
constant. Several attempts have been made to solve the problem
by two separate but simultaneous measurements. One of them
is to measure the moisture content by the one-parameter microwave method and the density by a gamma-ray attenuation
method [6]. Some other methods are based on measurement of
moisture content by two different frequencies, using microwave
and infrared ray or radio-frequency waves.

When the attenuation and phase shift of a penetrating
microwave are measured simultaneously, the moisture content
measurement can be independent of both the material layer
thickness and its density [4]. The measuring system is somewhat more expensive than that of the one-parameter system,
but it uses only a single sensor and it is fed from an oscillator
operating at a single frequency. The attenuation and phase
shift of a microwave passing through the sample of wet material
are related to the thickness of the sample and its dielectric
properties. Therefore, the attenuation and phase shift may be
expressed, respectively, as a function of the sample volume V,

mass of water M_w, and mass of dry materials M_d. The following
equation will be applied in most cases:

$$\frac{\phi}{L} = k_3\frac{M_w}{V} + k_4\frac{M_d}{V} \qquad (8.2.6)$$

where ϕ is the phase shift and k_3 and k_4 are the material
constants.

From Eqs. (8.2.5), (8.2.6), and (8.1.3b), the wet base
moisture content is represented as follows:

$$w_w = \frac{M_w}{M_d + M_w} = \frac{k_4 A - k_2\phi}{(k_1 - k_2)\phi - (k_3 - k_4)A} \qquad (8.2.7)$$

The dry base moisture content can be obtained by Eq. (8.1.3a).
The moisture content determined from Eq. (8.2.7) does not
depend on the thickness of the layer or the density of the
material. Figure 8.4 shows a block diagram of a two-parameter
measuring system [4].

The measurement of the moisture content of coal on a belt
conveyor showed that there were fluctuations in the attenuation
due to the belt [7]. The problem was overcome by use of an
integrator to cover the period of one belt revolution. Other
applications of microwave attenuation in iron ore sintering plants
will be found in the literature [8], including a block diagram of
a measuring system. A special measuring container can also be
applied in the moisture content measurement [9]. The material
flow can be decelerated in the container and the bulk density
could be stabilized. The container provides, at all times,
material in the measuring space. Short breaks in material supply
may cause rapid rises in the output signal, which might bring
the automatic control system out of adjustment. Further, temperature measurement could be done in the direct vicinity of the
measuring space. If the radiating and receiving antennas are
set in a measuring container with a metal shielding, a standing
wave will be produced in the container as in the case of the
microwave resonator discussed in Section 4.5.5. The effect is
less marked, however, when the measuring space is filled with
moist materials.

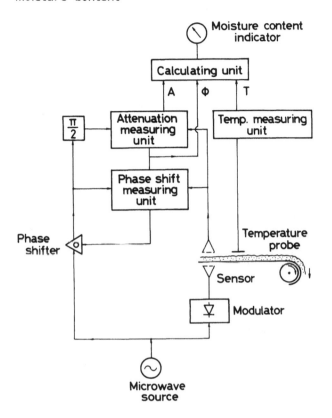

Figure 8.4 Two-parameter microwave moisture content measuring system.

8.3 REFLECTION OF INFRARED RAYS

Water absorbs infrared radiation through vibrations between the hydrogen and oxygen atoms at certain wavelengths. When the infrared ray is either reflected from or transmitted through a material, the degree of attenuation is a measure of the moisture content. For powders, a reflection moisture content meter is more suitable than the transmission type.

Figure 8.5 shows the transmissivity of infrared radiations through water as a function of wavelength. Relatively high

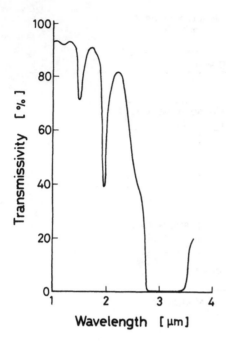

Figure 8.5   Transmissivity of infrared ray through water.

absorption occurs at 1.45, 1.93, and 2.95 μm. Among these
wavelengths, 2.95 μm is not suitable for the measurement because
the absorption is too high.   Small absorption is found at 1.2 μm,
but it is not enough.   An infrared ray of 1.93 μm wavelength
(water-absorbing ray) is commonly used.   When the moisture
content is extremely high, 1.45 μm is preferable.

The moisture content w_d is calculated by the following linear
relationship between w_d and attenuation A of the reflected ray:

$$w_d = a + bA \qquad\qquad (8.3.1)$$

where a and b are calibration constants. Attenuation A is given
by the following equation as in the case of the microwave moisture
meter:

$$A = -10 \log \frac{S}{I_0} \qquad\qquad (8.3.2)$$

where S is reflected power and I_0 is the incident power.

Figure 8.6 shows an outline of the infrared reflectance system. The infrared radiation through an optical filter is projected onto the moist materials by use of a mirror. Radiations reflected from the material surface are gathered by a concave mirror and the intensity is measured by use of a photocell such as PhS cell. Optical filters (interference filters) are set on a rotating disk.

The meter usually needs calibration for each material. In some cases, especially when the particles are hydrophobic, particle size affects the measured values. Dust and water vapor also affect the measurement. Therefore, the light path should be kept clean by use of, for example, air purging. As miscellaneous lights other than the measuring one also affect the measurement, they must be shielded. Further, optical systems are easily disturbed by vibrations and they need vibration-proof supporters. If the temperature of surrounding air is higher than 40°C, an air-cooling or water-cooling system is required so as to protect the optical system.

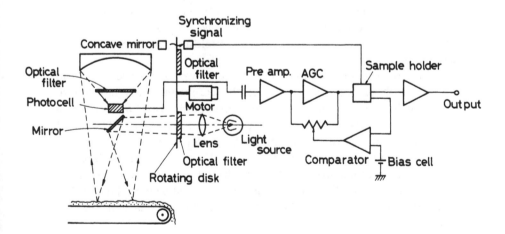

Figure 8.6 Infrared-reflectance moisture meter.

There are two types of infrared moisture meter: two-color
and three-color infrared moisture meters. They have two optical
filters and three optical filters on the rotating disk, respectively.
The two-color infrared moisture meter utilizes optical filters with
center wavelengths of 1.93 and 1.7 μm, which are sequentially
selected by the rotating disk [10]. The reflectance R_1 of a 1.7
μm-wavelength ray is not affected by water but is affected by
several factors, such as material temperature, color variations,
composition of particles, particle size, or surface irregularities.
These factors also affect the reflectance S of the 1.93 μm-wavelength ray as the disturbances of the moisture content measurements. The effects of these disturbances are effectively
eliminated through taking the ratio S/R_1 of reflected radiation
at these two wavelengths. That is, the infrared ray of 1.7 μm
works as a reference ray. Thus the incident power I_0 in Eq.
(8.3.2) is replaced by the reflectance R_1 of the reference ray.
This two-color method is based on the assumption that the
transmissivity of dry material at the water absorbing ray is the
same as that obtained at the reference ray. That is, the system
should have a horizontal base line for the transmissivity.
However, if the base line of the transmissivity is inclined by
some causes, as shown in Fig. 8.7, the reference reflectance R_1
is affected by the inclination r. Then the measured ratio will
be given by $S/(R_1 + r \, \Delta\lambda)$ instead of S/R_1, where $\Delta\lambda$ is the
difference of the two wavelengths. This will give an error for
the method of replacing the incident power I_0 by the reference
reflectance R_1. One of the methods avoiding the error is the
three-color method.

The three-color method [11] utilizes two reference rays and
the water-absorbing ray. One of the reference rays has longer
wavelength, $\lambda + \Delta\lambda$, than the water-absorbing ray, and the other
has shorter wavelength, $\lambda - \Delta\lambda$ (see also Fig. 8.7). The incident
power I_0 in Eq. (8.3.2) is replaced by $(R_1 + R_2)/2$, where R_2
is the reflectance at wavelength λ_2 ($= \lambda - \Delta\lambda$). The moisture
content calculations are carried out based on the ratio $2S/(R_1 + R_2)$. Then the inclination of the base line shown in Fig. 8.7
does not affect the ratio, because the error $r \, \Delta\lambda$ is canceled
with $-r \, \Delta\lambda$:

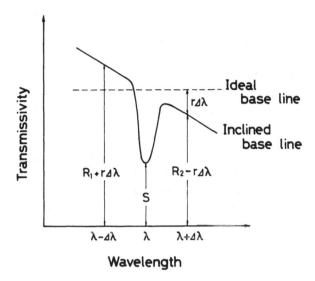

Figure 8.7   Effect of base-line inclination on the reflectance measurements.

$$\frac{2S}{(R_1 + r\,\Delta\lambda) + (R_2 - r\Delta\lambda)} = \frac{2S}{R_1 + R_2} \qquad (8.3.3)$$

Further, $R_1 + R_2$ may be modified to $cR_1 + dR_2$ by introducing additional parameters c and d which improve the calibration curves [12].   The parameters c and d are determined so as to increase the sensitivity of the meter under the condition of c + d = 1.

Another method to get reference reflectance is to change the wavelength of the reference ray so that maximum reflectance is obtained [13].   This modification improves the measurement of dark-color materials such as coal, where the usual fixed-wave-length method sometimes fails to detect the water content because of the weakness of the reflectance.

Infrared-reflection moisture meter does not need physical contact between the equipment and the test materials, and therefore it does not deteriorate or contaminate the materials. As is easily recognized by the principle, the measurement is based only on the surface powder, and the water content of

bulk powder cannot be obtained. If suitable plows or scrapers
are utilized, however, in order to make a fresh surface on the
optical head, the measured value will represent the water content
of the bulk powder. Such a system making a fresh surface also
has the purpose of maintaining a constant distance between the
optical head and the surface. If the distance is not enough,
changes in the distance caused by surface irregularities or other
factors, such as mechanical vibrations of the system, will affect
the measured reflectances. Although the distance from the
optical head to the material surface should be as large as possible,
the allowable distance now available is about 400 mm. The tested
area on the powder bed should also be as large as possible to
avoid the effects of a limited surface by averaging the reflectance
variations.

8.4 ELECTRICAL METHODS

The electric capacitance and electric resistance methods are
discussed here. The electric resistance method is also called the
electric conductance method.

8.4.1 Electric Capacitance Method

As discussed in Section 8.2, dielectric properties of a material
are represented by the complex dielectric constant. The microwave method utilizes primarily the imaginary part (loss factor)
of the complex dielectric constant. The electric capacitance
method, however, utilizes the real part (dielectric constant).
Although the real part also depends on the frequency of the
electromagnetic wave, as shown in Fig. 8.1 for water, it takes a
constant value (about 81 for water) at low frequency. Table
8.2 shows such a dielectric constant for various materials. As
the electric capacitance C of parallel-plate electrodes packed
with particles between them is given by Eq. (4.5.7), the dielectric
constant of the particles is obtained by the following equation:

$$\varepsilon = \frac{Cd}{A} \qquad\qquad (8.4.1)$$

Table 8.2   Dielectric Constant of Several Materials

Material	Dielectric constant (unitless)	Material	Dielectric constant (unitless)
Air	1	Glass	3.7–10
CO_2	1	Pottery	5–7
Water	80.7	Timber	2–6
Cyclohexane[a]	2.023	Epoxy resin	2.5–6
Benzene[a]	2.284	Acrylic resin	2.7–4.5
Chlorobenzene[a]	5.708	Cereals	3–5
Acetone[a]	21.3	Salt	6
Nitrobenzene[a]	35.7	Sugar	3

[a]Standard liquid for the measurement of dielectric constant.

where d is the distance between the parallel plates and A is the area of the electrode. The parallel-plate electrodes are made from circular disks or rectangular plates. In any case, one of the electrodes is smaller than the other. The smaller electrode is called the main electrode. The periphery of the main electrode is enclosed by an electrically conducting annular plate with insulating material between them to avoid the end effect of the capacitor.

If a capacitor of coaxial cylindrical electrodes is utilized, the dielectric constant is determined by the following equation [see also Eq. (6.3.1)]:

$$\varepsilon = \frac{C \ln(b/a)}{2\pi h} \tag{8.4.2}$$

where a is the outside diameter of the inner electrode of the capacitor, b the inside diameter of the outer electrode, and h the height of the cylinder constituting the main electrode.

The dielectric constant thus determined is a function of the water content, packing density of particles, and the chemical composition, as discussed in Section 4.5.3. Further, it depends

on the modes of water contained (bound water or unbound water).
Therefore, the relationship between the water content and the
dielectric constant of moist hygroscopic powder is different from
that of hydrophobic powder [14].

   Figure 8.8 shows three examples of the electric circuit for
measuring the unknown capacitance C of the powder bed.   The
method shown in Fig. 8.8a is called the frequency modulating
method, where the frequency of the oscillator output is given by
the following equation:

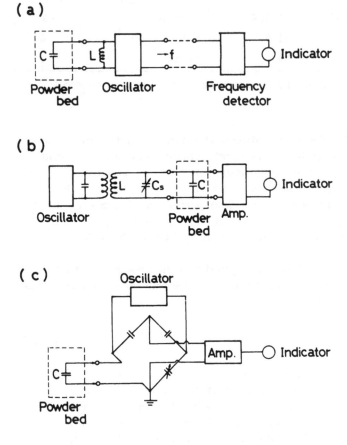

Figure 8.8   Electric circuits for the measurement of unknown
capacitance.

$$f = \frac{1}{2\pi} \frac{1}{\sqrt{LC}} \tag{8.4.3}$$

where L is the inductance of the coil. The oscillator output is
transmitted to a frequency detector, and then the capacitance is
calculated by Eq. (8.4.3). The output signal can be transmitted
a fairly long distance by this method. The temperature of the
oscillator should, however, be controlled in order to keep the
frequency constant.

The method shown in Fig. 8.8b is called the resonance method.
The oscillator generates an alternative current with known
frequency f by use of, for example, a quartz oscillator. The
unknown capacitance C is obtained by the following equation:

$$C = \frac{1}{(2\pi f)^2 L} - C_s \tag{8.4.4}$$

where C_s is the capacitance of the variable capacitor at which
the output current takes maximum value because of the resonance.
The value of the capacitance C_s can also be determined by a
tuning circuit.

The method shown in Fig. 8.8c utilizes an ac bridge. One
of the four capacitance elements is the powder bed. Electric
current caused by the bridge unbalance is utilized to measure
the capacitance of the powder bed.

The frequency f of the oscillator output is selected so as
to detect the water content effectively. In some cases, a frequency as high as 50 MHz will be utilized. The dielectric constant
of water, in this frequency range, does not depend on the
frequency (see also Fig. 8.1). In an electrical sense, the powder
bed should be represented by an RC parallel element instead of
the single C element. However, the unknown resistance R does
not affect the capacitance measurement as far as the resistance
is high enough.

Figure 8.9 shows an example of the moisture content meter
adopted in a powder handling process [15]. Particles are sampled
from the main line into the measuring line. First the sampled
mass is measured by use of a load cell. And then the capacitance
of the powder bed is obtained by a cylindrical capacitor. After

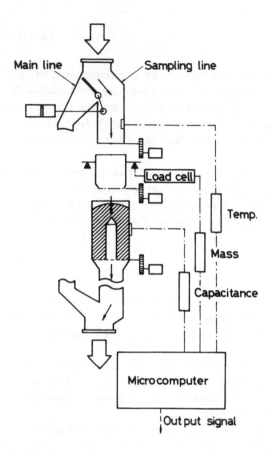

Figure 8.9 Moisture content meter in a powder handling process.

finishing these measurements, the sampled powder is returned
into the main line. The temperature of the powder is also
measured. The data on the sampled mass, the capacitance, and
the temperature are sent to a microcomputer and the moisture
content is calculated based on a calibration curve.

8.4.2 Electric Resistance Method

Electric resistance of a wet powder bed decreases with increasing
moisture content. As in the case of capacitance measurements,

parallel-plate electrodes or cylindrical electrodes are utilized. Sampled powder is packed between the electrodes by use of a similar sampling system, as shown in Fig. 8.9. Then the current I through the powder bed is measured under a constant applied voltage V. The so-called specific resistance ρ is more adequate than the simple resistance R in order to represent the electrical property of the powder. For the parallel-plate electrodes, it is given by the following equation:

$$\rho = \frac{VA}{Id} = R\frac{A}{d} \qquad\qquad (8.4.5)$$

As is clear from Eq. (8.4.5), the specific resistance has the dimensions of Ω m. For the coaxial cylindrical electrodes, it is given by the following equation:

$$\rho = \frac{2\pi h}{\ln(b/a)}\, R \qquad\qquad (8.4.6)$$

The specific resistance does not depend on the electrode constants: A, d, h, a, and b. The typical relationship between the specific resistance ρ and the moisture content w_d is as follows:

$$\ln\,\rho = -k_1 w_d + k_2 \qquad\qquad (8.4.7)$$

where k_1 and k_2 are constants determined through calibration. The specific resistance also depends on temperature, modes of water contained, packing density of particles, and the chemical composition. Higher temperature gives lower specific resistance, because both the concentration and the mobility of charge carriers increase with temperature. Usually, the temperature of the powder bed is measured simultaneously by use of a thermocouple in order to correct the effect. If the specific resistance is very high, as in the case of dried plastic particles, it takes a long time until the measured resistance shows the steady value. This is because the electric charge is stored between the particles. The measurement is not adequate for on-line applications.

If the water is kept on the surface of each particle, the powder bed shows very low resistance even if the moisture content is not as high. Therefore, the specific resistance for hydrophobic particles decreases very sharply as the moisture content increases. If hygroscopic particles are treated by adding water, the initial reading indicates a much higher moisture content than that of the tempered particles because of the change in modes of water contained [16].

Figure 8.10 shows the typical electric circuits for the measurement of electric resistance. A dc power supply is preferably utilized because it is stable. Batteries can also be used as the dc power supply. It should, however, be taken into consideration that measurement error will be caused by the

(a)

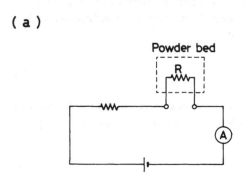

(b)

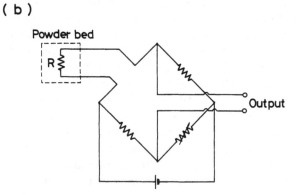

Figure 8.10   Electric circuits for measurement of unknown resistance.

electrolysis of water. In this case, the current will gradually
decrease. If an ac power supply is applied, the effect of the
electrolysis can be depressed. Also, signal processing, such as
signal amplification, becomes easier. But the measuring system
can be unstable because of the lead-wire capacitances. Figure
8.10b shows the electrical bridge for measuring the electric
resistance of the powder bed. Bridge unbalance is easily detected.
The resistance value obtained by balancing the bridge does not
depend on the voltage of the electric power supply. Usually,
the current output caused by the unbalance is utilized in calculation of the moisture content based on a calibration curve.

The resistance measurements depend largely on the degree
of contact between particles and the electrodes, which is not as
important in the capacitance measurements. The electrodes for
the resistance measurements are therefore specially designed to
fit them to the sampled particles.

8.5 NMR AND OTHER METHODS

The methods for measuring moisture content discussed here are
based on nuclear magnetic resonance (NMR), neutron reflection,
cobalt chloride reaction, or equilibrium relative humidity.

8.5.1 Nuclear Magnetic Resonance

When a material is placed in a uniform dc magnetic field, precession of nuclei having definite magnetic moment results. The
sampled material is set in a steady magnetic field as shown in
Fig. 8.11. The coil around the sample produces a radio-frequency
magnetic field which is perpendicular to the steady magnetic field.
If the frequency of the alternating magnetic field coincides with
that of the precession of the nuclei, the nuclei resonate in these
fields, causing radio-frequency energy to be absorbed. The
condition for resonance is given by the following equation:

$$2\pi\upsilon = \gamma H_0 \qquad\qquad\qquad (8.5.1)$$

where υ is the radio frequency, γ the gyromagnetic ratio of the
nuclei, and H_0 the strength of the main dc magnetic field. The

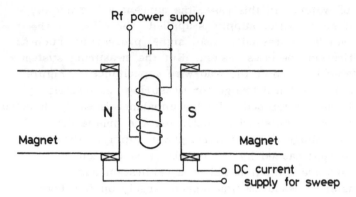

Figure 8.11 NMR spectrometer.

frequency at which the resonance occurs is called the Lamor
frequency. The gyromagnetic ratio divided by 2π is 4257.7 Hz/G
(gauss) for hydrogen. Therefore, the hydrogen resonance would
occur at about 4.26 MHz in a field H_0 of 1 kG. The behavior
in the neighborhood of resonance can be obtained by sweeping
either the frequency of the exciting RF (radio frequency)
-magnetic field or the strength of the dc magnetic field. A set
of coils wound around the permanent magnet sweeps the field
strength over a limited range by adjusting the current in the
coils.

The energy absorbed by the resonance can be detected as
a change in the impedance of the electric circuit, including the
RF coil. The RF coil works as both the RF magnetic field coil
and the receiver coil. The precessing nuclei induce a weak
voltage superimposed on the RF voltage. The coil voltage is
amplified and detected. The output of the detector is recorded
against the sweep of the dc field. Thus the absorption curve
is obtained. Further, the derivative curve (Fig. 8.12) will be
obtained through signal modulation. The resonance lines in the
absorption or the derivative curve are broad for hydrogen in
water tightly bound to a surface, while they are narrow for
hydrogen in unbound water [17]. The peak-to-peak amplitude
of the derivative curve is not affected by the broad line, and a

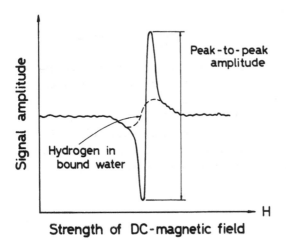

Figure 8.12   Derivative curve of NMR absorption.

good correlation can be obtained between the peak-to-peak amplitude and the moisture content of a starch sample [18].   The calibration curve is not a straight line because of both the variation of line width with moisture content and the possible variation in the ratio of bound water to unbound water.

The moisture content of a continuous stream of coal can also be measured by feeding the coal into the NMR spectrometer [19]. The sample container used in the literature was made from a Teflon tube on which a copper wire was wound.   The wire works as the RF coil.   The moist sample was fed into the Teflon tube from a conical hopper above the spectrometer.   The flow of coal under gravity was regulated at 100 g/h by means of a worm-gear drive at the base of the sample container.   The peak-to-peak height of the derivative signal was successfully taken as a measure of the moisture content.   If the sample flow velocity is too high, the signal output will decrease, as in the case of measurement of light-water concentration in mixtures of light (H_2O) and heavy (D_2O) water [20].

8.5.2 Neutron Reflection

Neutron moisture meters utilize the so-called fast neutrons produced by a nuclear source such as ^{241}Am-Be (americium-beryllium). If a neutron collides with a nuclei in the sample material, it loses kinetic energy and the neutron is decelerated. The energy loss is represented by $(E_2/E_1)_g$, where E_1 and E_2 are the energy of neutron before and after the collision, respectively and $(E_2/E_1)_g$ represents the geometric mean of E_2/E_1. For collision with protons (hydrogen, mass number = 1), the energy loss is given by the following equation:

$$\left(\frac{E_2}{E_1}\right)_g = \frac{1}{e} \tag{8.5.2}$$

Now, the logarithmic attenuation ratio ξ is defined by

$$\xi = -\ln\left(\frac{E_2}{E_1}\right)_g \tag{8.5.3}$$

and $\xi = 1$ for the collision with proton. The logarithmic attenuation ratio ξ decreased with increasing nucleus mass number. For example, it is 0.136 for ^{14}N amd 0.035 for ^{56}Fe. The actual energy loss of the neutron also depends on the scattering cross section σ_s of the target nucleus. The degree of neutron deceleration by a nucleus is proportional to the product $\xi\sigma_s$. If the product $\xi\sigma_s$ is assumed to be 100 for proton, it is 0.452 for ^{14}N and 0.038 for ^{56}Fe. Therefore, collision between the neutron and the hydrogen nucleus (proton) causes the greatest energy loss.

The fast neutron thus loses its energy and changes to a thermal neutron, which is reflected by the powder bed. The energy of the thermal neutron is about 0.025 eV (electron volts) at room temperature (it is given by kT, where k is Boltzmann's constant). The number of thermal neutrons reflected is detected by use of a boron-trifluoride counter (BF_3 counter) based on the following nuclear reaction:

$$^{10}B + n \rightarrow {}^{7}Li + \alpha \tag{8.5.4}$$

where n represents the thermal neutron and α denotes the α-ray.
The boron-trifluoride gas is ionized by the produced α-ray, and
therefore current pulse will be produced.

As discussed above, the number of thermal neutrons reflected
is almost solely determined by the amount of hydrogen in the
sample material. That is, the neutron moisture meter measures
not the moisture content but the amount of hydrogen. Therefore,
the bound water is not distinguished from the unbound water.
The amount of chemically combined hydrogen in the material
should also be taken into consideration in the measurement of
moisture content [21].

Figure 8.13 shows a detection head of the neutron moisture
meter. Such a detection head is used either by inserting it into
a powder bed, by setting it on the wall of a hopper, or by
putting it on a powder bed. When it is used to measure the
moisture content of a powder bed on a belt conveyor, a stream-
forming unit described in Section 8.2 should be utilized.

The moisture content obtained by the neutron reflection
method is volumetric. That is the mass of water contained in
the unit volume of powder bed. To convert volumetric moisture
content to a mass basis, it must be divided by the bulk density
of the powder bed. The measurement of the bulk density can be
done based on the gamma-ray attenuation method described in
Section 5.4 [21,22]. The effective volume of the sample measured
by the moisture meter depends on the chemical components and
the moisture content of the material. For low-moisture material,

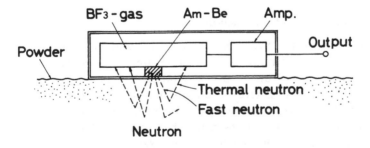

Figure 8.13 Neutron moisture meter.

the measured volume is larger than that for high-moisture material, because neutrons can reach longer distance. If some of the fast neutrons pass through the sample, the number of the thermal neutrons detected will decrease. Therefore, the effective sample volume should be full of powder, especially in the low-moisture case. A penetration neutron moisture meter has also been studied to improve the measurement accuracy in the low-moisture range [23]. The measurements also depend on the location of water in the powder bed. The water near the meter head can decelerate the fast neutrons more effectively [21].

8.5.3 Cobalt Chloride Reaction

Ethanol solution of cobalt chloride reacts with moisture, and the color of the solution changes from light blue to pink. The color change can be detected by use of a colorimeter. The colorimeter output is then related to the moisture content through calibration.

The measurements will be carried out by use of a sampling system. Particles are continuously sampled from the main line in powder handling processes by use of a special feeder. The sampled particles are mixed with 0.1% $CoCl_2$ ethanol solution in a vessel by use of a magnetic stirrer. Then the cobalt solution is filtered off and pumped to the colorimeter. Although materials that form unsoluble compounds with cobalt or chloride ions will affect the measurements, the reaction is highly specific for moisture [22].

8.5.4 Equilibrium Relative Humidity

As discussed in Section 8.1, the moisture content of a powder tends to equilibrium with the surrounding air. Conversely, the relative humidity of air near the powder bed is approximately in equilibrium with the moisture content of the powder bed. Therefore, if the relative humidity of air near the powder bed is measured, the moisture content will be estimated based on a calibration curve for each material. For example, the equilibrium moisture content at 60% relative humidity is 0.5% for asbestos,

1% for kaolin, and 14% for flour. Although there are many types of humidity meter, the difference between the temperature of the powder bed and the wet-bulb temperature will also be utilized in the measurement [24].

REFERENCES

1. W. L. McCabe and J. C. Smith, Unit Operations of Chemical Engineering, McGraw-Hill Book Company, New York, p. 885 (1956).

2. R. Toei (ed.), Kanso Sochi (Dryer), 8th ed., Nikkan Kogyo, Tokyo, p. 24 (1974).

3. L. H. Busker, Instrum. Control Syst. 41, 89 (1968).

4. A. Kraszewski, J. Microwave Power 15, 209 (1980).

5. T. Makimoto and S. Matsuo, Maikroha Kogaku No Kiso (Fundamentals of Microwave Technology), Hirokawa Publishing Co., Tokyo, Chap. 12 (1979).

6. M. Tiuri, K. Jokela, and S. Heikkilae, J. Microwave Power 15, 251 (1980).

7. D. A. Hall, J. C. Sproson, and W. A. Gray, J. Inst. Fuel 45, 163 (1972).

8. T. Siraiwa, S. Kobayashi, A. Koyama, M. Tokuda, and S. Koizumi, J. Microwave Power 15, 255 (1980).

9. A. Kraszewski, S. Kulinski, J. Madziar, and K. Zielkowski, J. Microwave Power 15, 267 (1980).

10. K. Hoffman, Chem. Ing. Technol. 35, 55 (1963).

11. K. Miyauchi, Keiryo Kanri 33, 397 (1984).

12. Y. Nambu, Keiso 27(12), 24 (1984).

13. Y. Terada and N. Takanabe, Keiso 28(2), 53 (1985).

14. H. Nakajima, T. Kuramae, and T. Tanaka, Funsai 17, 3 (1972).

15. Y. Yamada and M. Yasuguchi, Sensor Gijutu 5(12), 62 (1985).

16. A. A. Abdelrahman and E. P. Farrell, Cereal Chem. 58, 307 (1981).

17. D. W. Jones, Chem. Process Eng. 47, 22 (1966).

18. A. R. Aikman, R. S. Codrington, and F. F. Kirchner, Control Eng. 4(6), 105 (1957).

19. A. R. Lander and A. E. Stacey, Br. J. Appl. Phys. 13, 136 (1962).

20. A. M. J. Mitchell and G. Phillips, Br. J. Appl.Phys. 7, 67 (1956).

21. Y. Inoue, M. Sakaki, and Y. Imaru, Toshiba Rev. 21, 1126 (1966).

22. R. Milton, Chem. Eng. 76(16), 83 (1966).

23. Y. Amano, N. Ohkubo, H. Nishikawa, H. Tominaga, N. Wada, N. Tachikawa, Y. Kuramochi, S. Horiuchi, and Y. Satoh, Tetsu To Hagane S-65 (1982).

24. F. C. Harbert, Special POWTEC 73 issue, p. 45 (1973).

9

Pressure

The subjects discussed here are primarily concerned with the measurement of powder pressure. Load cells, which are widely utilized in hopper scales or belt scales, are discussed in this chapter. Measurement of fluid pressure is also discussed because some technical considerations are involved when particles are suspended in the fluid. In Section 9.1 we outline powder pressure and its measurement.

9.1 INTRODUCTION

Powder pressure is a normal component of stress exerted by particles. The force acting on the wall of a powder storage vessel is not always normal to the wall, even if the powder in the vessel is stationary. Therefore, there may be a stress component parallel to the wall surface, which is called the frictional stress (shear stress). The powder pressure in a storage vessel does not show a linear increase with the powder depth, but tends to a saturation pressure. This characteristic is quite different from that for fluids, where the pressure is directly proportional to the depth. For example, the vertical pressure P_V exerted by particles stored in a cylindrical vessel will be given by Janssen's equation:

$$P_v = \frac{D \rho_B \underline{g}}{4 \mu_w k} \left[1 - \exp\left(-\frac{4 \mu_w k}{D} h \right) \right] \tag{9.1.1}$$

where D is the diameter of the cylindrical vessel, ρ_B the bulk density of powder, μ_w the wall frictional coefficient, k the ratio between the horizontal pressure and the vertical pressure, and h the depth of the powder bed. The term $D\rho_B \underline{g}/4 \mu_w k$ gives the saturation pressure. The horizontal pressure P_h acting normal to the wall surface is given by kP_v. The pressure ratio k (= P_h/P_v) is called Janssen's or Rankine's coefficient.

Equation (9.1.1) is derived based on a simple force balance. If the particles are flowing down in the storage vessel, the powder pressure takes a different value from that obtained by the equation. Powder pressure acting on the wall may take a fairly high value in such a dynamic state. Pressure gauges set on the wall should therefore have enough strength against the highest pressure caused by the powder flow. Also, the pressure receiving unit of the pressure gauge should be stiff enough. The contact between particles and the surface of the pressure receiving unit is not so uniform as in the case of fluids. The powder pressure may act as only a small part of the surface area because of the discrete nature of particles. The pressure gauge should, even in these situations, give the same value irrespective of the place where the pressure is acting. Further, it is required to have high resistance against abrasion caused by hard particles.

The powder pressure causes a small displacement of the pressure receiving unit through deformation. The displacement is then transformed into an electric signal. The transformation is carried out based on various physical phenomena described in Section 9.3. The displacement of the pressure receiving unit should be given by an elastic deformation so as to avoid the hysteresis in the transformation of the pressure into the mechanical displacement.

9.2 PRESSURE GAUGE CONSTRUCTION

The pressure receiving unit of a pressure gauge is usually
constructed by use of a steel diaphragm, as in the case of a
level switch shown in Fig. 6.4. Pressure gauges of this type
are widely used in powder pressure measurements. However, it
is preferable to introduce a special configuration called parallel-
plate construction in the pressure receiving unit [1,2]. Figure
9.1 shows the principle of parallel-plate construction. The head
of the pressure receiving unit is constructed by use of a rigid

a)

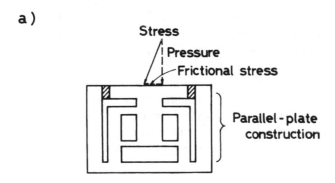

b)

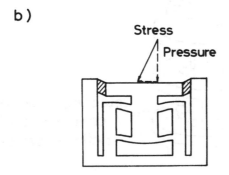

Figure 9.1   Parallel-plate construction.

steel block, and therefore, it does not deform. The deformation by the applied stress takes place at the parallel-plate construction in the pressure receiving unit, as shown in Fig. 9.1b. Further, only the normal component (pressure) of the stress is effective to the deformation, because the deformation of the pressure receiving unit is confined to the direction normal to the parallel plates.

The frictional stress will also be detected by the use of parallel-plate construction [1]. Figure 9.2 shows the detecting block by which the x component and y component of the frictional stress are obtained separately. The detecting block (called the stress cell) has two pairs of parallel-plate construction set at mutually orthogonal positions. Three pairs of parallel-plate construction may compose a stress cell for detecting all three components of a stress. Thus the stress acting on a wall can be identified perfectly. Even though the system may become complicated, various modifications have been carried out in the design of stress cells. For example, drilled holes connected to each other are utilized instead of parallel plates. Biaxial wall stress cells following the parallel-plate principle has been applied to the continuous and simultaneous measurements of the powder pressure and the frictional stress acting on the wall of a storage vessel [3].

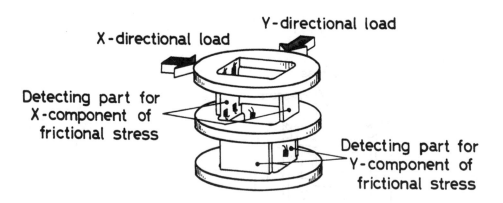

Figure 9.2   Biaxial wall stress cell.

9.3   PRESSURE TRANSDUCERS

9.3.1   Strain Gauges

Strain gauges are most widely utilized in pressure transducers. The electric resistance of a conductive material is proportional to its length L and inversely proportional to the cross-sectional area A:

$$R = \rho_0 \frac{L}{A} \tag{9.3.1}$$

where R is the electric resistance and ρ_0 is the specific resistance of the material.* If the material is stretched, the resistance increases. Therefore, the deformation of the strain gauge can be transformed into an electric signal. Strain gauges are made from a conductive material which has relatively high resistivity, such as Constantan (Cu-Ni alloy). They are formed into a special shape as shown in Fig. 9.3. The wire strain gauge is formed by use of a small-diameter wire (e. g., 0.025 mm in diameter), while the foil strain gauge is made through etching. The foil strain gauge has a rectangular cross section, and higher electric current is admissible than for wire strain gauge. The base of a strain gauge is usually made from Japanese paper or plastic.

If a small strain is applied to a wire, the electric resistance of the wire will be changed according to Eq. (9.3.1). The resistance increment ΔR should therefore satisfy the following equation:

$$\frac{\Delta R}{R} = \frac{\Delta L}{L} - \frac{\Delta A}{A} + \frac{\Delta \rho_0}{\rho_0} \tag{9.3.2}$$

The last term in Eq. (9.3.2) can be neglected for an isotropic conducting material. Further, if the volume V = AL is not changed by the strain, the following equation will be obtained:

$$\frac{\Delta A}{A} = - \frac{\Delta L}{L} \tag{9.3.3a}$$

---

*The specific resistance for an isotropic conducting material is assumed to be constant.

a)   b)

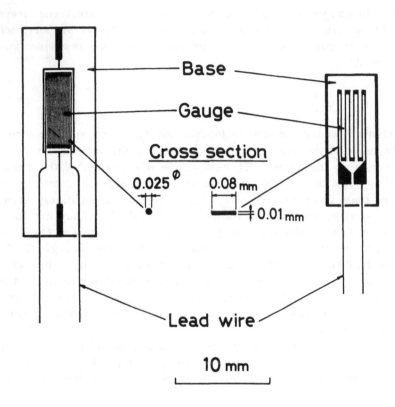

Figure 9.3 Strain gauges: (a) wire type; (b) foil type.

Substitution of Eq. (9.3.3a) into Eq. (9.3.2) gives the following equation for an isotropic conducting material with no volume change:

$$\frac{\Delta R}{R} = 2 \frac{\Delta L}{L} = 2\varepsilon \qquad\qquad (9.3.4a)$$

where ε is the strain ($= \Delta L/L$). The more general form of Eq. (9.3.3a) is given by

$$\frac{\Delta A}{A} = -2\nu \frac{\Delta L}{L} \qquad\qquad (9.3.3b)$$

where ν is Poisson's ratio. For most substances, Poisson's ratio is between 0.2 and 0.4, usually 0.3. From Eq. (9.3.3b), the more general form of Eq. (9.3.4a) is obtained as follows:

$$\frac{\Delta R}{R} = \left[(1 + 2\nu) + \frac{\Delta \rho_0 / \rho_0}{\varepsilon} \right] \varepsilon \qquad\qquad (9.3.4b)$$

The actual strain gauges have complicated configurations as shown in Fig. 9.3, and, in practical applications, Eqs. (9.3.4a) and (9.3.4b) are replaced by the following equation:

$$\frac{\Delta R}{R} = F \varepsilon \qquad\qquad (9.3.5)$$

F in Eq. (9.3.5) is called the gauge factor or, in other words, Eq. (9.3.5) defines the gauge factor of the strain gauge.

The variation of the electric resistance is measured by use of an electric bridge shown in Fig. 9.4. In this figure, R_1 represents the resistance of the strain gauge and R_2 to R_4 are known resistances. The output voltage of the bridge is given by the following equation:

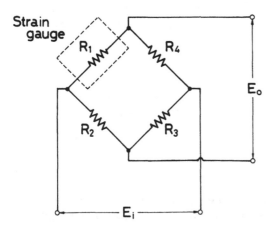

Figure 9.4 Electrical bridge detecting resistance increment.

$$E_o = \frac{R_1 R_3 - R_2 R_4}{R_1 R_2 + R_1 R_3 + R_3 R_4 + R_2 R_4} E_i \qquad (9.3.6)$$

The bridge will be kept in balance ($E_o = 0$) by setting the resistances so as to satisfy the relation $R_1 R_3 = R_2 R_4$. If the resistance of the strain gauge is increased to $R_1 + \Delta R$, the bridge will be out of balance. The output voltage for a small value of $\Delta R / R_1$ is given by the following equation:

$$E_o = \frac{\Delta R / R_1}{2 + (R_2/R_3) + (R_3/R_2)} E_i \qquad (9.3.7)$$

From Eqs. (9.3.5) and (9.3.7), it can be seen that the output voltage E_o is directly proportional to the strain ε.

Various electric bridges are developed to compensate the effects of temperature variations or additional resistances of lead wires [4]. Temperature increase will cause the expansion of the wire gauge resulting in a voltage unbalance, even though no stress is applied. Thermoelectromotive force between the resistance elements and lead wires also affects the balance. Further, if an alternative current is applied to the electric bridges, it is also required to compensate the effects of capacitances between the resistance elements and the surroundings.

The dynamic characteristics of these strain gauges will depend on the sound velocity transferred in the gauges. The sound velocity v in a solid material is given by the following equation:

$$v = \sqrt{\frac{E}{\rho}} \qquad (9.3.8)$$

where E is the elastic coefficient of the material and ρ is the density. The highest frequency that can be detected by the strain gauge is estimated by

$$f = \frac{v}{kL} \qquad (9.3.9)$$

where the factor k is usually taken to be about 10. If the length of the wire gauge L is 1 cm and the velocity v is 4000 m/s, the frequency f is estimated as 40 kHz.

Figure 9.5 shows a load cell which is utilized in belt scales or hopper scales, discussed in Section 4.2. Strain gauges are fixed on the steel rod by use of adhesives such as thermosetting resin. They are usually covered entirely by a shielding case. Inert gases will sometimes be utilized to protect strain gauges from moisture attack. The diameter of the rod where strain gauges are attached is made a little smaller than the remaining part. This tends to avoid the stress concentration which causes a nonuniform stress distribution in the rod. Several types of load cells are developed, as shown in Fig. 9.6, according to the various requirements in the measurements.

For a diaphragm pressure transducer, strain gauges are set at the reverse side of the diaphragm. There is a special transducer which has a secondary small diaphragm behind a main large one and the space between the two diaphragms is filled with mercury [5]. Strain gauges in this transducer are attached at the reverse side of the secondary diaphragm. A small deformation of the main diaphragm pushes the mercury and the deformation is transferred to the secondary diaphragm after it is amplified. In this system, the main diaphragm can be made from harder metal than the normal uni-diaphragm pressure gauge, because the smaller deformation can also be detected by the secondary diaphragm through amplification. As discussed in Section 9.2,

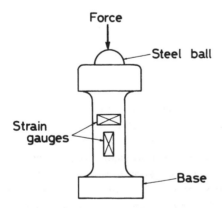

Figure 9.5   Load cell.

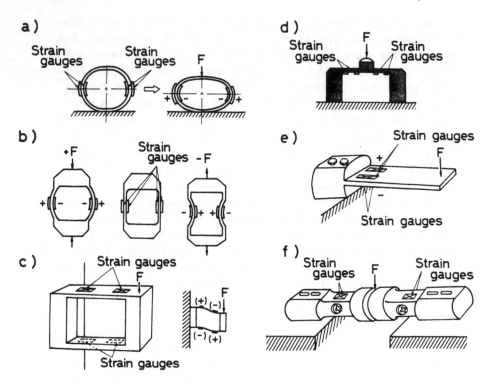

Figure 9.6   Various types of load cells.

however, parallel-plate construction is preferable for the measurement of powder pressure.

9.3.2   Differential Transformers

Some pressure transducers utilize differential transformers so
as to detect the deformations or strains.   Figure 9.7a is a
schematic diagram showing the principle of the differential transformer.   As in the case of an electric transformer, two types
of coils (called the primary coil and the secondary coil, respectively) are formed on a cylindrical bobbin.   A plunger connected
to, for example, a diaphragm is inserted into the core space of

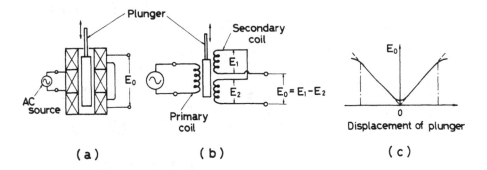

Figure 9.7   Differential transformer.

the bobbin.   The secondary coil is composed of two identical coils
connected differentially, as shown in Fig. 9.7b.   If ac current
is supplied to the primary coil, voltage output E_0 will be obtained
through the secondary coil according to electromagnetic induction.

$$E_o = - M \frac{dI_1}{dt} - L_2 \frac{dI_2}{dt} \tag{9.3.10}$$

where I_1 is the current supplied to the primary coil, I_2 the
induced current in the secondary coil, M the mutual inductance
of the coils, and L_2 the self-inductance of the scondary coil.
Both the mutual inductance and the self-inductance are directly
proportional to the magnetic permeability of the core space.
      However, the secondary coil of the differential transformer
produces no output voltage as long as the magnetic permeability
of the core space is maintained constant, because the secondary
coil is composed of two identical coils connected differentially and
is in balance.   If the permeability is partly changed by a plunger
displacement, the secondary coil loses the electric balance.   Then
the output ac voltage or ac current is obtained from the secondary coil as shown in Fig. 9.7c.   The output voltage is directly
proportional to the displacement of the plunger from 1 μm to
about 500 μm.
      Figure 9.8 shows a pressure transducer of this type [5].
The differential transformer is set in the central part.   The

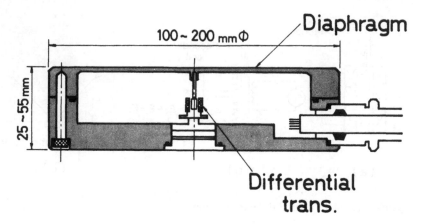

Figure 9.8 Pressure transducer (differential transformer type).

plunger is connected to the diaphragm, and the coils are fixed
on the base. Therefore, the diaphragm and the differential
transformer are mechanically independent. Because of this
system, the differential transformer has a high stability for long
use. Further, the output signal is large enough to operate a
recorder or a process controller without amplifiers. The temperature effect is much smaller than that for strain gauges and
temperature compensation is not required. Protection against
moisture effects is also easier than for strain gauges. The
frequency of the ac source, however, should be carefully controlled because the output voltage E_o is dependent on the current
frequency, as is easily recognized from Eq. (9.3.10). The
capacitance of a transfer cable will also affect the output signal.

9.3.3 Semiconductor Gauges

If a semiconducting crystal such as a single crystal of silicon is
deformed by an applied stress, the concentration of charge carrier
in the crystal will be greatly changed. It is well known that the
conductivity σ of such a semiconductor is given by the following
equation:

$$\sigma = e(\mu_n n + \mu_p p) \qquad\qquad (9.3.11)$$

where e is the elementary charge (1.6×10^{-19} C), n the concentration of free electrons, p the concentration of positive holes,
μ_n the mobility of electrons, and μ_p the mobility of positive
holes. As the specific resistance ρ_0 is given by the inverse of
the conductivity, it is greatly changed by the deformation
(piezoresistance effect). For such a semiconducting crystal,
the electric resistance increase ΔR is determined solely by the
specific resistance increase $\Delta\rho_0$:

$$\frac{\Delta R}{R} = \frac{\Delta\rho_0}{\rho_0} = F\varepsilon \qquad (9.3.12)$$

The gauge factor F is 100 to 200 for semiconducting crystals,
while it is about 2 for the wire strain gauges, as discussed in
Section 9.3.1. A small diaphragm or a strain gauge is made
from a single crystal of silicon and the surface is treated by
impurity (usually boron for a p-type semiconductor) diffusion to
control the concentration of charge carriers. The sensor is as
small as a few millimeters square. Therefore, very small, high-
sensitivity pressure transducers can be constructed [6,7].
However, the temperature effect is larger than that for wire
strain gauges, because the concentration of charge carriers is
dependent on the temperature. For example, the concentration
of free electrons is given by the following equation:

$$n = N \exp\left(\frac{E_f}{kT}\right) \qquad (9.3.13)$$

where E_f is the Fermi level discussed in Section 4.5.1, k is
Boltzmann's constant, T the temperature, and N the proportional
constant. Therefore, the additional temperature-compensating
gauge should be placed near the semiconductor strain gauge.
It is also possible to make a temperature-compensating circuit
on the single crystal of silicon by use of the microelectronic
technique [8]. Further, an integrated circuit including the
linearizing circuit, signal processing circuit, and so on, can be
constructed on the same crystal.

9.3.4   Electromagnetic Load Cells and Others

Electromagnetic Load Cells

If an electrically magnetized ferromagnetic material is deformed
by an applied stress, the density of the magnetic flux through
the electric coil will change.   This phenomenon is known as the
inverse magnetostrictive effect or Villari effect.   The deformation
of the ferromagnetic material should be elastic so as to avoid the
hysteresis.   Figure 9.9 is the schematic diagram of the electromagnetic load cell (magnecell, Yasukawa Denki Seisakusho, Co.,
Ltd.).   The output voltage is obtained as a difference between
the induced voltage and the dc bias voltage.   Load cells of this
type are strong against the unexpected impulsive forces.

Carlson Pressure Transducers

The pressure transducers of Carlson type are based on the same
principle as the wire strain gauges discussed in Section 9.3.1,
but they utilize strong piano wires [5].   The piano wires are
stretched or loosened by use of a mechanical system constructed
behind a diaphragm, and the electric resistance increase or
decrease is detected by use of an electric bridge as discussed in

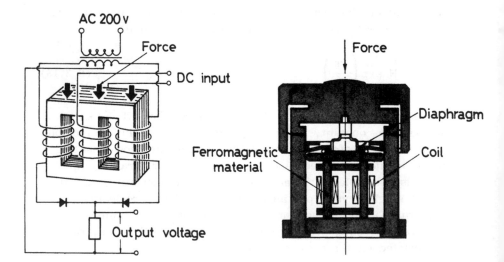

Figure 9.9   Electromagnetic load cell.

Section 9.3.1. Pressure transducers of this type have higher durability than that of wire or foil strain gauges.

Electric Capacitance Pressure Sensors

As discussed in Section 8.4.1, the electric capacitance of a parallel-plate condenser is inversely proportional to the gap between the two plates. If one of the plates forms a pressure receiving diaphragm, its displacement caused by pressure can be detected as a capacitance increase. The diaphragm is made from fused quartz and an electric conductive material is coated on it so as to form an electrode of the capacitor. The gap between the two plates is filled with silicone oil. The electric capacitance is measured by an electric circuit such as that discussed in Section 8.4.1.

Quartz Crystal Oscillator as a Pressure Sensor

The quartz oscillator discussed in Section 5.3.7 is also applied to the measurement of pressure. If the quartz crystal is stressed, the frequency of the oscillation will be shifted as in the case of the quartz crystal microbalance. As the digital output is obtained by this sensor, it easily interfaces with telemetry and digital computer systems [8,9].

Hall Element as a Pressure Sensor

If a Hall element is displaced in mutually opposite magnetic fields as shown in Fig. 9.10, dc voltage proportional to the displacement will be obtained. A dc current is supplied to the Hall element through a pair of electrodes. The charge carriers in the Hall element suffer the electromagnetic force. As in the case of the electromagnetic flowmeter discussed in Section 4.5.4, a potential difference will be formed in the direction perpendicular to both the current supplied and the magnetic field. If the Hall element is in balancing position, the potential difference E_1 caused by the left-hand-side magnetic field is canceled by the potential difference E_2 caused by the right-hand-side field. The output voltage is null in this case. If the element is moved from the balancing position, the voltage output $E_1 - E_2$ directly proportional to the displacement will be obtained. The Hall element is made from a semiconducting material such as germanium.

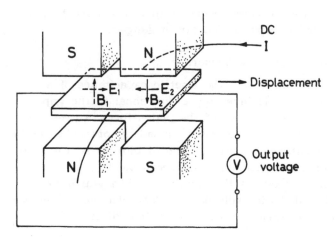

Figure 9.10   Hall element as a pressure sensor (displacement is detected).

9.4   PRESSURE MEASUREMENT IN GAS-SOLID SUSPENSION FLOW

Pressure drop caused by a flowing gas-solid suspension has been discussed in Section 4.4. Measurement of the pressure drop is an important subject not only for the mass flow measurement discussed there but also for the estimation or control of the energy consumption in powder handling processes. The energy consumption is directly proportional to the pressure drop multiplied by the gas flow rate.

Pressure sensors utilizing diaphragms have been applied to liquid-solid suspension flow, but diaphragms have suffered from erosion and corrosion. Further, particles deposited on them caused considerable error in the measurement. Although pressure sensors utilizing fiber optics are under development [8], existing sensors need some considerations against the intrusion of fine particles into the pressure taps. These particles will deposit on the inside wall, and finally, the taps will be blocked. Further, the measurement of fluid dynamic pressure by a Pitot tube will

be affected by fine particles, because the kinetic energy of a
particle changes into the fluid pressure during particle deceleration in the pressure probe.

9.4.1 Protecting Pressure Taps by Air Purge

The air purging system is shown in Fig. 9.11. The system is
composed of a compressor, air purging lines, air flowmeters,
pressure chambers, and pressure taps. The purging air velocity should be maintained higher than the particle deposition
velocity discussed in Section 3.2.3. The air pressure p^* in the
chamber will be measured instead of the pressure p at the wall
of the main duct. There may be some difference between these
pressure values [10]:

$$p^* - p = \Delta p_f + \Delta p_c \tag{9.4.1}$$

where Δp_f is the pressure drop caused by the purging flow
between the chamber and the main duct, and Δp_c is the pressure
drop caused by the confluence between the purging flow and
the main flow. The pressure drop Δp_f can be calculated by use
of the Fanning's equation. On the other hand, the pressure

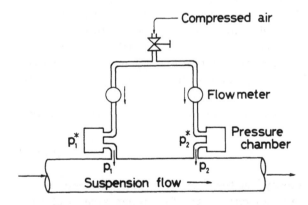

Figure 9.11 Air purging system protecting pressure taps from
particle intrusion.

drop Δp_c depends on the velocity ratio u_m/u_p, where u_m is the average velocity of the main flow and u_p is that of the purging flow. Therefore, measurement of the static pressure needs an experimental relationship between the pressure drop Δp_c and the velocity ratio [10]. However, these pressure drops do not affect the measurement of the pressure difference between two points along the main duct, because they take almost the same value for each of the purging systems. That is,

$$p_1^* - p_2^* = p_1 - p_2 \tag{9.4.2}$$

The air purging system is fairly complicated, because the flow rate of purge air should be kept constant. A simpler method is to set a piano wire in the center of a pressure tap. If the piano wire is protruded a little into the main flow, it may vibrate itself and protect the pressure tap from the blockage [11]. A piezoelectric or electromagnetic oscillator may also be utilized for this purpose. A mechanical cleaner is another possibility for tap protection.

9.4.2 Conversion of Particle Momentum into Pressure

Pitot tubes are widely utilized in the measurement of gas velocity. If particles are suspended in the flow, some of them will intrude into the Pitot tube. These particles are decelerated in the tube, because the gas in the tube is stagnant. That is, the particles entering the tube lose their kinetic energy in the tube, which will cause a pressure increase [12]. If a particle entering the tube collides with the wall, the kinetic energy is partly changed into another type of energy, such as heat or energy for plastic deformation.

The pressure increase Δp is given by the following equation:

$$\Delta p = \frac{1}{2} \rho m u_0^2 \zeta \tag{9.4.3}$$

where ρ is the air density, m the mass flow ratio defined by Eq. (4.4.1), u_0 the air velocity at the far upper stream of the Pitot tube, and ζ the conversion ratio of the particle kinetic

energy into the pressure. The conversion ratio ζ is calculated
by the following equation:

$$\zeta = 1 - \frac{\int_0^{r_c} 2\pi r (v/u_0)^2 \, dr}{\pi R^2 \eta} \tag{9.4.4}$$

where η is the particle collection efficiency in the tube, v the
particle velocity just before the collision, R the inner radius
of the Pitot tube, and r_c the radius of a circular area at the
far upper stream such that all the particles passed through the
area enter the Pitot tube. The particle collection efficiency is
a monotone increasing function of the inertia parameter introduced
in Section 4.4 [13]. As discussed in Section 3.2, the inertia
parameter for a particle smaller than 10 μm is represented as
follows:

$$\psi = \frac{C_m \rho_p D_p^2 u_0}{18 \mu D} \tag{9.4.5}$$

where C_m is Cunningham's slip correction factor, given by Eq.
(3.2.12).

The particle velocity v in Eq. (9.4.4) is also a monotone
increasing function of the inertia parameter, because particles
having larger inertia are less decelerated. Figure 9.12b shows
the calculated results on the pressure conversion ratio as a
function of the inertia parameter [14]. As shown in Fig. 9.12a,
a wire mesh is assumed to be set at position x in the tube. If
a particle intruding into the tube is blocked by the mesh, the
particle kinetic energy will be absorbed by the mesh, resulting
in a smaller conversion ratio. Parameter X in the figure is a
dimensionless position of the mesh. As the mesh is nearer the
entrance, the conversion ratio becomes smaller and the pressure
measurement error caused by the particles becomes smaller. The
inertialess particle gives the largest effect on the pressure
measurement ($\zeta = 1$). In this case, the measured pressure is
equal to that for a heavy gas with a density of $\rho(1 + m)$.

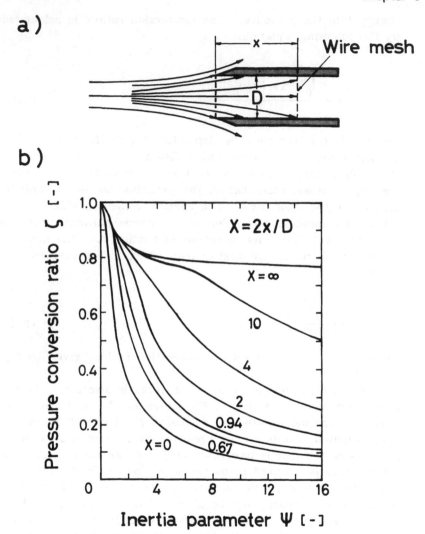

Figure 9.12 Pressure conversion ratio as a function of inertia parameter for various positions of wire mesh.

REFERENCES

1. K. Chijiiwa, Y. Hatamura, K. Ono, and T. Mizutani, Nippon Kikai Gakkai Ronbunshu 48, 1767 (1982).

2. K. Iinoya (ed.), Funtai Kogaku Binran (Powder Technology Handbook), Nikkan Kogyo, Tokyo, p. 770 (1986).

3. T. Takeuchi, T. Nagao, Y. Hatamura, and N. Nakajima, J. Soc. Powder Technol. Jpn. 21, 406 (1984).

4. T. Aoyagi, Wire Strain Gauges (Japanese), Nikkan Kogyo, Tokyo, p. 105 (1963).

5. K. Iinoya (ed.), Funtai Keisoku Handbook (Handbook for Instrumentation of Powder Handling Processes), Nikkan Kogyo, Tokyo, Chap. 24 (1981).

6. T. Jotaki and R. Moriyama, J. Res. Assoc. Powder Technol. Jpn. 10, 386 (1973).

7. Y. Kato, Toyota Gihou Jpn. 10(2), 1 (1969).

8. H. Idegawa and A. Ishii, in (Kogyo Chosakai ed.) Sensor Katsuyo Gijutsu (Applications of Sensors), Kogyo Chosakai, Tokyo, p. 244 (1984).

9. J. M. Paros, ISA AC, p. 602 (1972).

10. K. Iinoya, N. Tanaka, and T. Kawase, J. Res. Assoc. Powder Technol. Jpn. 5, 1193 (1968).

11. R. G. Boothroyd, Flowing Gas-Solids Suspension, Chapman & Hall Ltd., London, p. 95 (1971).

12. J. L. Dussourd and A. H. Shapiro, Jet Propul. 28, 24 (1958).

13. H. Yoshida, H. Masuda, and K. Iinoya, J. Assoc. Powder Technol. Jpn. 17, 117 (1980).

14. S. Yuu, Y. Uchiyama, T. Yokoyama, and K. Iinoya, Kagaku Kogaku 38, 237 (1974).

10

Temperature

Temperature measurement is required in the control of several types of equipment, such as a cement kiln, spray dryer, or powder calcinator. The temperature sensors are classified in two groups: thermal contact and radiation thermometers.

10.1 THERMAL CONTACT THERMOMETERS

The temperature of powder stored in a vessel will be measured by use of a thermal contact thermometer. Thermocouples, electric-resistance thermometers, thermistors, and transistor thermometers are included in this category.

10.1.1 Thermocouples

Thermocouples are composed of two different types of metal wires connected at both ends, making a loop. One of the ends is called the hot junction (sensing point) and the other the reference

junction (or cold junction). The hot junction is usually welded.
If there is a temperature difference between these two junctions,
electromotive force (thermoelectromotive force) will be generated
in the loop. The electromotive force V is a function of the
temperature difference Δt and the reference temperature t_0 as
follows:

$$V = a \, \Delta t + b[2t_0 \, \Delta t + (\Delta t)^2] \qquad\qquad (10.1.1)$$

where a and b are dependent on the type of metal wires. The
phenomenon is called the Seebeck effect. The second term on
the right-hand side of Eq. (10.1.1) is usually neglected and the
electromotive force generated is, approximately, directly proportional to the temperature difference. Table 10.1 shows the common
thermocouples, their composition, and the common operating-
temperature range. Iron-constantan thermocouples are suitable
to reducing atmospheres, chromel-alumel thermocouples to oxidizing atmospheres, and copper-constantan thermocouples to low-
temperature atmospheres. Copper-constantan thermocouples are
preferable where moisture is high. Platinum-platinum rhodium
thermocouples can be used in either oxidizing or reducing
atmospheres. The values of the electromotive force for these
thermocouples are found in the standard calibration tables [1].

In practice, the extension lead wires can be used between
the two junctions as shown in Fig. 10.1. The terminals are
connected to a millivoltmeter or a potentiometer. The measurement system should be electrically shielded because the output
voltage is very small. The reference junction is usually kept
0°C by use of an ice bath. An electrical cooler or electrical
compensator will also be utilized, so that the reference junction
appears to be held at a constant temperature.

The sensing parts of the thermocouples are protected against
corrosive action or mechanical damage by use of ceramic tubes
or metal tubes filled with alumina or magnesium oxide powder.
The electrical insulation of thermocouples is also made by use of
these protecting tubes. Thermocouples covered with protecting
tubes are available as small as 0.2 mm in diameter.

Table 10.1 Thermocouples

Materials	Abbr.	Temp. range (°C)	EMF (µV/°C)	Accuracy (°C)
Copper: constantan	CC	-180–300	50	2–5
Iron: constantan	IC	0–600	60	3–10
Chromel: alumel	CA	0–1000	40	2–10
Platinum: platinum 10% rhodium	PR	100–1400	10	0.5–5

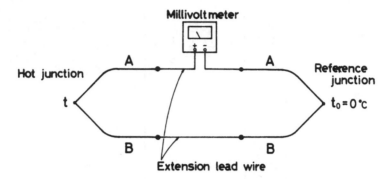

A,B: different metallic wire

Figure 10.1   Thermoelectric circuit with extension lead wires (thermocouple).

10.1.2   Resistance Thermometers

Electric-resistance thermometers utilize the characteristic of metals that the electric resistance of conductors (metals) increases with an increase in ambient temperature.   The electric resistance of semiconducting materials also changes when the ambient temperature is changed.   In the latter case, the concentration and mobility of the charge carrier in materials increase with increasing temperature T (K), and therefore the resistance R (Ω) decreases. The temperature dependence of the electric resistance of semiconducting materials is given by the following equation:

$$R = aT^{-b} \exp\left(\frac{c}{T}\right) \qquad\qquad (10.1.2)$$

where a, b, and c are calibration constants.   Resistance thermometers constructed of semiconducting materials are known as thermistors (thermally sensitive resistors).   Table 10.2 shows the types of resistance thermometers and their operating temperature ranges.

Industrial resistance thermometers are usually constructed of platinum.   The platinum wire (diameter between 0.03 and 0.05 mm and length 1 m) is wound on a mica plate or on a glass rod

Table 10.2   Resistance Thermometers

Materials	Temperature range (°C)	Accuracy (%)
Platinum	-180—500	0.5—2
Nickel	-50—120	0.5—2
Copper	0—120	0.5—2
Thermistor	-50—200	0.5—2

and is inserted into a protecting tube or enclosed in tempered glass.   Figure 10.2 shows the platinum resistance thermometer enclosed in glass [2].   The electric resistance of platinum resistance thermometers changes 0.4% for a temperature change of 1°C (temperature coefficient = 0.4%/°C).   The temperature coefficient of thermistors is about 10 times larger than that of the platinum resistance thermometer.   The temperature-resistance characteristic of thermistors is, however, nonlinear, as is represented by Eq. (10.1.2), and the calibration constants a, b, and c will be different for each thermistor.   Although thermistors have such a defect, their sensing elements are very small (0.3 to 2 mm in diameter) and the dynamic response can be faster (the time constant is 0.3 to 2 s) than that of platinum resistance thermometers.

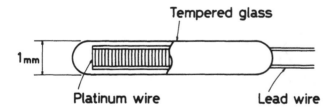

Figure 10.2   Platinum resistance thermometer.

10.1.3 Transistor Thermometers and Others

A silicon transistor can be utilized as a thermometer. NPN
transistors, for example, are consisted of three small blocks, as
shown in Fig. 10.3a. A block of p-type semiconductor (charge
carrier = positive hole) is sandwiched by two blocks of n-type
semiconductors (charge carrier = electron). Each block has lead
wires, called base, emitter, and collector, respectively. Figure
10.3b shows a symbolized NPN transistor.

   The silicon transistor that is utilized as a thermometer has
a temperature coefficient of about -2 mV/°C for the voltage
between the base and the emitter. The voltage V_{BE} is represented by the following equation[2]:

$$V_{BE} = V_{g0} - \frac{kT}{e} \ln \frac{aT^r}{I_C} \qquad\qquad (10.1.3)$$

where V_{g0}, a, and r are constants, k is Boltzmann's constant

a)

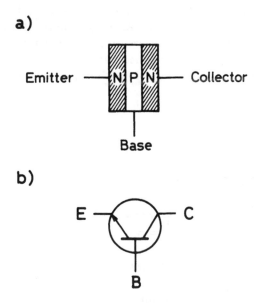

Figure 10.3 NPN transistor.

$(1.38 \times 10^{-23}$ J/K), T (K) the temperature, e the elementary
charge, and I_C the current supplied to the collector.

The thermal energy kT in Eq. (10.1.3) is about 0.025 eV
at 20°C. The base-emitter voltage V_{BE} is usually between 0.2
and 1.2 V, while the collector current is between 10 and 100
µA. Equation (10.1.2) will be applied in the temperature range
−50 to 150°C, where the voltage is nearly proportional to the
temperature. As in the case of the thermistors, the characteristics of the transistor thermometers are not same for all transistors.
By changing the collector current I_C, however, it is easily
calibrated to have the same thermometer characteristics.

Bourdon tube thermometers are also included in thermal
contact thermometers. The sensitive part (bulb) of a Bourdon
tube thermometer is filled with a gas or liquid and is connected
through a capillary. The gas or liquid in the bulb changes in
volume, pressure, or vapor pressure depending on the temperature. The change is transferred to the Bourdon tube through
the capillary. Then the Bourdon tube changes the shape uniquely
depending on the bulb temperature. The Bourdon motion can
be utilized to drive the pointer of a temperature indicator or
to drive a pen recorder. Fluid-filled bulbs deliver enough
power to drive controller mechanisms and even directly actuate
control valves without electric power sources. The distance of
remote indication is, however, at most 30 m and the temperature
measurements are less accurate.

There are other thermometers, such as bimetal thermometers
with optical fibers and optical fiber thermometers with light-
sensitive semiconductors, phosphor layers, or birefractive plates
[3]. Optical fiber thermometers with semiconductors utilize the
fact that the light intensity transmitted through semiconductors
(GaAs or CdTe) depends on the semiconductor temperature.
The accuracy is said to be ±0.5°C for measurements between
−30 and 300°C.

Thermal contact thermometers should satisfy the effective
thermal contact between the powder and the sensing part. There
should be as many particles in volume as 10 times the diameter
of the sensing part so as to attain enough contact. Temperature
measurements of particles suspended in gas or liquid are carried
out by sampling the particles in a probe in which thermocouples

are mounted. The measurements should be done in the process line. If the sampled powder is drawn out from the process line, heat loss causes measurement error. The heat capacity of the probe will also affect the particle temperature, and it should be kept as small as possible. After finishing the measurement, the sampled particles are purged into the main flow.

If the sensing points are far apart from the control room, as in the case of temperature measurements of powder in storage vessels, remote scanners with a two-wire circuit will be utilized [4]. Optical fibers will also be useful in such a case.

10.2 RADIATION THERMOMETERS

All objects emit thermal radiation energy depending on their temperature. Therefore, the temperature of particles can be estimated by detecting the thermal radiation energy. Radiation thermometers need no thermal contact with particles. Thermal radiation energy of a perfect emitter (blackbody) is given by Planck's law:

$$E(\lambda, T) = \frac{2\pi c^2 h}{\lambda^5} \left[\exp\left(\frac{c_2}{\lambda T}\right) - 1 \right]^{-1} \qquad (10.2.1)$$

where λ is the wavelength, h is Planck's constant (6.6256×10^{-27} erg·s), c the light velocity (2.99793×10^{10} cm/s), k is Boltzmann's constant (1.38054×10^{-16} erg/K), and c_2 = ch/k = 1.4387×10^4 μm·K.

Equation (10.2.1) shows that the radiation energy is concentrated mainly in the visible and infrared region of wavelength below 10 μm. From Eq. (10.2.1), the total energy is given by the following equation (the Stefan-Boltzmann law):

$$E(T) = \int_0^\infty E(\lambda, T) \ d\lambda = 4.88 \left(\frac{T}{100}\right)^4 \qquad \text{kcal/m}^2 \cdot \text{h} \qquad (10.2.2)$$

Equations (10.2.1) and (10.2.2) can be utilized in temperature measurements.

There are three types of definitions of solid materials temperature:

1. Radiation temperature, T_R. If the emissive power of an object is equal to that of a perfect emitter of temperature T_R, T_R is called the radiation temperature of the object.
2. Luminance temperature, T_S. If the monochromatic luminance of an object is equal to that of a perfect emitter of temperature T_S, T_S is called the luminance temperature of the object.
3. Color temperature, T_F. If the color of an object is the same as that of a perfect emitter of temperature T_F, T_F is called the color temperature of the object.

There is the following relationship between the luminance temperature T_S and the true temperature T:

$$\frac{1}{T} = \frac{1}{T_S} - \frac{\lambda}{c_2} \ln \varepsilon_\lambda \tag{10.2.3}$$

where ε_λ is monochromatic emissivity (blackness) of the object. Further, the relationship between the color temperature and the true temperature is given by the following equation:

$$\frac{1}{T} = \frac{1}{T_F} + \frac{1}{c_2} \frac{\ln(\varepsilon_{\lambda_1}/\varepsilon_{\lambda_2})}{(1/\lambda_2) - (1/\lambda_1)} \tag{10.2.4}$$

where $\varepsilon_{\lambda 1}$ and $\varepsilon_{\lambda 2}$ are emissivities at the wavelengths of λ_1 and λ_2, respectively. From Eqs. (10.2.3) and (10.2.4), the following relationship will be derived:

$$T_S < T < T_F \tag{10.2.5}$$

The emissivity of a powder bed is usually approximated by unity (blackbody) and therefore, $T \cong T_S \cong T_F$. This is not true for a particle or suspended particles.

Table 10.3 shows the common thermosensor (detector), measuring temperature ranges and the accuracy of radiation thermometers. Radiation pyrometers measure the total energy $E(T)$. The radiation is gathered by lenses (optical condenser) and focused

Table 10.3   Radiation Thermometers

Thermometer	Detector	Temperature range (°C)	Accuracy (%)
Radiation	Thermopile	200--1000	±1−2
pyrometer	Silicon cell	600−3000	±0.5−1
	Thermistor bolometer	0−500	±0.5−1
	PbSe, PbS	100−1000	±0.5−1
Two-color	PbS	300−1000	±0.5−1
pyrometer	Silicon cell	700−2000	±0.5−1
	Photomultiplier	1000−3500	±0.5−1

on a detector such as a thermopile, silicon cell, thermistor bolometer, or photocell (PbSe, PbS). Thermopiles are composed of thermocouples connected in series so as to increase the sensitivity. The accuracy and stability are, however, not as good as that of other detectors. Portable radiation pyrometers are also available.

Two-color pyrometers are more expensive than radiation pyrometers. They have, however, several features. Figure 10.4 shows the principle of two-color pyrometers. The radiation from particles is condensed by lenses and projected to an inter-ference filter made from, for example, indium phosphate (InP). The radiation reflected by the filter and that transmitted through the filter are measured by photocells, respectively, and the ratio of the output signals of each photocell are calculated. As the wavelength of reflected radiation is different from that of transmitted radiation, the luminance temperature at two different wavelengths (two colors) is thus obtained. The measured temp-erature is insensitive to variations in the emissivity of particles, and is less affected by several disturbances caused by aerosols such as water vapor or fumes. Two different optical filters may also be utilized in order to get radiation intensities at two different wavelengths.

A two-wavelength pyrometer can also be applied to gas-solid suspension flow [5]. The radiation from particles is passed

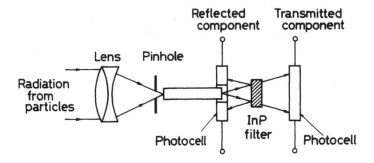

Figure 10.4   Principle of two-color pyrometers.

through suitable pairs of monochromatic filters inserted in dia-
metrical opposition into a metal disk rotating at speeds up to
8000 rpm.   The two alternating beams are focused on the cathode
of a photoelectric cell, and the resulting voltages are measured
by use of an oscilloscope.   This instrument was used in connection with a study of the rate of reduction of iron oxide spherical
particles of 5 to 100 μm in diameter conveyed in a stream of
hydrogen flowing downward in an atomized-suspension reactor at
reactor wall temperatures ranging from 500 to 1100°C.

A more economic thermometer for the measurement of temperature ranging from 700 to 3000°C is an optical pyrometer,
which is a kind of brightness pyrometer.   The pyrometer determines
the temperature of an object by comparing the luminance of the
object with that of a reference filament of a standard electric
bulb.   The acuracy of optical pyrometers is about ±5°C at 1000°C.
It is easy to control the electric current supplied to the filament
so as to get the same luminance as the object.   Such a pyrometer
is called an automatic optical pyrometer.

The measuring points of radiation thermometers can be changed
by use of a rotating or vibrating mirror.   These instruments,
called scanning thermometers, are used to obtain the temperature
profile of large equipment such as a cement kiln.   Commercially
available infrared scanning thermometers can be applied in the
measurement of a temperature range of 10 to 1000°C with a

sensitivity of 0.5°C at 25°C and a resolving power of 3% of full
scale. The focal length of the instruments is between 1.2 and
600 m. An infrared temperature profile radiometer can even be
utilized in measurement of the earth's surface from a spacecraft
[6].

In these measurements, the effect of emissivity of the
objective materials on the measured temperature should be taken
into consideration. Further, if there is heavy dust in the light
path, the radiation may be attenuated as discussed in Section
5.4. An air purge will be used in these cases. Optical fibers
are also utilized [3], especially where radiation thermometers
are difficult to be set because of obstacles in the light path.
Optical fiber thermometers also have the advantage of not being
affected by electrical noise. Water cooling must sometimes be
carried out when the thermometers become too hot. It should,
however, be noted that water condensation will take place on
the optical parts when the temperature becomes too low. Then
the measurements become erroneous.

REFERENCES

1. R. C. West and M. J. Astle (eds.), CRC Handbook of
    Chemistry and Physics, 61st ed., CRC Press, Inc., Boca
    Raton, Fla., p. E-108 (1980).

2. H. Yamasaki (ed.), Jido Seigyo Handbook—Kiki Ouyou Hen
    (Handbook for Automatic Control—Instruments & Applications),
    Ohmsha Ltd., Tokyo, Chap. 1 (1983).

3. K. Morita, Keiryo Kanri 33, 91 (1984).

4. K. Iinoya (ed.), Funryuutai Keisoku Handbook (Handbook
    for Instrumentation of Powder Handling Processes, Nikkan
    Kogyo, Tokyo, p. 442 (1981).

5. N. J. Themelis and W. H. Gauvin, Can. J. Chem. Eng. 40,
    157 (1962).

6. G. S. Bowen, ISA Trans. 13, 101 (1974).

11

Viscosity of Slurries, Powder Composition, and Motion Sensors

This chapter is concerned primarily with the on-line viscosity measurement of slurries. An optical method for detecting the mixture quality of powders is also introduced. Further, so-called motion sensors applicable to powder feeders are discussed briefly.

11.1 SLURRY VISCOSITY

11.1.1 Introduction

The viscosity of slurries depends on various factors, such as particle volume fraction in a slurry, size distribution of suspended particles, particle shape, and electric charge on particles. If a shear stress is applied to a liquid, flow will be induced in the direction of the stress. It is well known that liquid water satisfies the following relationship between the shear stress τ (Pa) and the flow velocity u (m/s):

$$\tau = -\mu \frac{du}{dy} \qquad (11.1.1)$$

where μ (Pa·s) is the viscosity of water and y (m) is the
coordinate taken normal to the flow.   The velocity gradient,
−du/dy (s^{-1}), is called the shear rate.   Equation (11.1.1) also
shows the fact that shear stress will be caused whenever the
velocity gradient is formed in the fluid.   Fluids that satisfy
Eq. (11.1.1) are called Newtonian fluids.   All gases and homogeneous nonpolymeric liquids are included in Newtonian fluids.
Any fluid that does not satisfy Eq. (11.1.1) is referred to as
a non-Newtonian fluid.   For example, a coal water mixture (CWM),
aqueous nuclear slurries, and water suspensions of clays are all
included in the category of non-Newtonian fluids.   For non-
Newtonian fluids, apparent viscosity μ_a is defined by the following equation:

$$\mu_a = -\frac{\tau}{du/dy} \tag{11.1.2}$$

where the apparent viscosity μ_a may depend on either the shear
rate −du/dy or the shear stress τ.   The apparent viscosity is
also called the consistency.   Numerous empirical equations have
been proposed to express the relationship between the shear
stress and the shear rate [1].

The apparent viscosity for dilute suspension of spherical
particles is represented by the following theoretical equation [2]:

$$\mu_a = \mu(1 + 2.5\phi) \tag{11.1.3}$$

where φ (unitless) is the volume fraction of particles and μ is
the viscosity of the liquid vehicle.   Equation (11.1.3) is called
Einstein's equation.

For slurries of higher particle concentration, several theoretical or semiempirical equations have been proposed to represent
the apparent viscosity [3]: for example, Mooney's equation [4],

$$\mu_a = \mu \exp\frac{2.5\phi}{1 - k\phi} \quad \text{(for } \phi < 0.5) \tag{11.1.4}$$

where the parameter k = 1 ~ 1.5, and Mori-Ototake's equation [5],

$$\mu_a = \mu\left[1 + \frac{D_{p50}S_v\phi}{2(1 - \phi/\phi_c)}\right] \tag{11.1.5}$$

wher D_{p50} (m) is the mass median diameter of particles, S_v (m^{-1}) the specific surface area on a volume base, and ϕ_c the maximum volume fraction of particles obtainable for the system.

The following equation may also be applied to high-concentration slurries [6]:

$$\mu_a = \mu \, \exp\left\{\left[2.5 + \left(\frac{\phi}{\phi_c - \phi}\right)^{0.48}\right]\frac{\phi}{\phi_c}\right\} \qquad (11.1.6)$$

11.1.2 Viscosity Sensors

The following discussion is concerned with the instruments that may be applied to on-line measurements of the apparent viscosity. For laboratory measurements of viscosity, several types of instruments are available [7].

Capillary Viscometers

The wall shear stress τ_w exerted by fluids flowing in a straight circular tube is given by the following equation*:

$$\tau_w = \frac{\Delta p}{2L} R \qquad (11.1.7)$$

where Δp is the pressure drop, R the inside radius of the tube, and L the length of the tube. On the other hand, the velocity gradient $-du/dy$ on the wall is given by the following equation:

$$-\frac{du}{dy} = \frac{1}{\pi R^3}\left[3Q + \Delta p \, \frac{dQ}{d(\Delta p)}\right] \qquad (11.1.8)$$

where Q is the volumetric flow rate of the fluid. Equation (11.1.8) can be derived by taking a momentum balance on the flowing fluids (Rabinowitsch equation [8]). From these two equations, the relationship between the shear rate and the shear stress can be obtained for any fluid as long as the flow is laminar and

---

*From the force balance, the following equation will be obtained:

$$\pi R^2 \, \Delta p = 2\pi R L \tau_w$$

isothermal. Therefore, the apparent viscosity can also be obtained through measurements of the pressure drop Δp and the volumetric flow rate Q.

In practical viscosity measurements, however, the Hagen-Poiseuille equation for Newtonian fluids is usually applied:

$$\mu_a = \frac{\pi \Delta p R^4}{8LQ} \qquad\qquad (11.1.9)$$

Figure 11.1 shows a schematic diagram of a capillary viscometer. The slurry is sampled from the process line to the measuring system by use of a constant-rate pump. As the viscosity is easily affected by the temperature, the measuring system is set in a constant-temperature bath and the sampled slurry is warmed up to the bath temperature by use of a heat exchanger. Then the slurry flows through a narrow tube (capillary). The pressure drop caused by the narrow tube is detected by use of a pressure transducer such as those discussed in Section 9.3. The differential pressure signal gives the viscosity data through calculations based on Eq. (11.1.9). Measurements of flow rate under constant-pressure drop can also be applied in the viscosity measurement. The value of the viscosity obtained by these

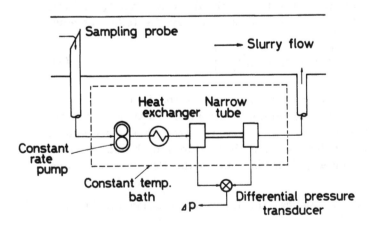

Figure 11.1 Capillary viscometer.

methods is, however, not the same as those obtained by other methods described below, except for the Newtonian fluid, because the method based on the Hagen-Poiseuille equation is used only for convenience.

Falling-ball Viscometers

Figure 11.2 shows a schematic diagram of a falling-ball viscometer. The slurry is sampled from the process line to the measuring system by use of a suitable pump. The measuring system should be set in a constant-temperature bath as in the case of the capillary viscometer. The sampled slurry flows through the measuring vertical cylinder, where a small steel sphere is suspended. For the viscosity measurement, the flow is suddenly changed to the bypass line by use of a three-way magnetic valve. Then the steel sphere falls down in the cylinder. The time elapsed during the fall is measured by use of a timer and magnetic sensors which are set along the measuring cylinder with a known distance between them. After the measurement, the flow is changed again to the measuring cylinder and the steel sphere is lifted up.

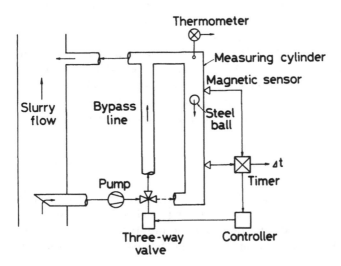

Figure 11.2   Falling-ball viscometer.

The equation of motion for the sphere of diameter D_p is given by the following equation, similar to Eq. (7.2.3), accounting for the bouyancy force caused by the fluid:

$$m_p \frac{dv}{dt} = 3\pi\mu_a D_p(u - v) + m_p g\left(1 - \frac{\rho}{\rho_p}\right) \tag{11.1.10}$$

where ρ is the density of the slurry. At the steady state of falling, the time derivative of the velocity $dv/dt = 0$. As the fluid velocity $u = 0$, the settling velocity is given by the following equation:

$$v_{st} = \frac{(\rho_p - \rho)D_p^2 g}{18\mu_a} \tag{11.1.11}$$

Equation (11.1.11) is called the Stokes equation, and the apparent viscosity determined based on the equation is called the Stokes viscosity. As discussed in Section 7.2, the particle (sphere) Reynolds number should be kept small enough ($Re_p < 2$). For larger Reynolds number, modifications will be necessary in the fluid drag term. Also, in this case, a longer distance is required before the measuring section.

Further, there may be a considerable wall effect on the settling velocity, because the fluid flow around the sphere will be restricted by the wall of the measuring cylinder. The fluid drag on the sphere will be increased by the wall effect, and the measured settling velocity is lower than that obtained by Eq. (11.1.11). The retardation effect on the settling velocity can be represented by the following equation when the creeping flow around the sphere is assumed (Ladenburg's equation) [7,9]:

$$v_m = \frac{1}{1 + 2.1(D_p/D)} v_{st} \tag{11.1.12}$$

where v_m is the measured settling velocity and D is the diameter of the measuring cylinder. Although Eq. (11.1.12) is valid for $D_p/D < 0.06$, extended equations correcting the wall effect are also available [7,10]. The effect of the bottom of the measuring cylinder has also been studied [11]. The effect can, however, easily be avoided by setting the measuring section with enough distance from the bottom.

In the case of the falling-ball viscometer, shear stress and shear rate cannot be calculated because the flow around the sphere is complicated compared with the capillary viscometer. In these situations, only the maximum shear rate is estimated as $3v_m/D_p$ around the surface periphery at the maximum cross-sectional diameter.

When only a relative value of viscosity is desired, the following simplified form of Eq. (11.1.11) will be utilized:

$$\mu_a = k(\rho_p - \rho)\, \Delta t \qquad\qquad\qquad (11.1.13)$$

where k is the calibration constant and Δt is the time of fall. As for the reference value of the viscosity, 1.002×10^{-3} Pa·s of distilled water at 20°C and 101.325 kPa (atmospheric pressure) may be utilized. More viscous liquids can also be available as reference materials from several research centers (e.g., National Research Laboratory of Metrology, Japan; Cannon Instrument Company, United States).

Coaxial Cylinder Viscometers (Rotational Viscometer)

Figure 11.3 shows a coaxial cylinder viscometer. A rotating cylinder is coaxially set in an outer cylinder with a small gap between them. The sampled slurry is very slowly pumped into the viscometer. After the viscosity is measured, it is returned to the main line. The inner cylinder is rotating at a constant rotational speed and the torque acting on the shaft is detected by use of a torquemeter which is constructed with strain gauges (discussed in Section 9.3.1) and slip rings for transferring the electric signals from the strain gauges attached on the rotating shaft to the outside detecting electrical circuit.

The torque T acting on the shaft is given by the following equation:

$$T = (\text{area}) \times (\text{radius}) \times (\text{shear stress})$$

$$= 2\pi R_a H R_a \tau_a \qquad\qquad\qquad (11.1.14)$$

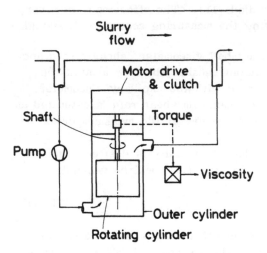

Figure 11.3 Coaxial cylinder viscometer.

where R_a is the radius of the rotating cylinder, H the height of the cylinder, and τ_a the shear stress at the outer surface of the inner cylinder.

If it is assumed that the fluid is Newtonian, the tangential laminar flow velocity u between the two vertical coaxial cylinders is represented by the following equation (radial and axial flows are assumed to be negligible) [11]:

$$u = \frac{R_a^2 \omega}{R_b^2 - R_a^2} \frac{R_b^2 - r^2}{r} \qquad (11.1.15)$$

where r is the radial coordinate and ω is the angular velocity of the rotating cylinder. By use of Eq. (11.1.15) and the stress equation in cylindrical coordinates [11], the shear stress τ is obtained as follows:

$$\tau = -\mu r \frac{d}{dr}\left(\frac{u}{r}\right)$$

$$= 2\mu \frac{R_a^2 R_b^2 \omega}{R_b^2 - R_a^2} \frac{1}{r^2} \tag{11.1.16}$$

where μ is the viscosity of the Newtonian fluid. Replacing μ by μ_a and substituting R_a into r, the shear stress τ_a for the viscometer is obtained as follows:

$$\tau_a = 2\mu_a \frac{R_b^2 \omega}{R_b^2 - R_a^2} \tag{11.1.17}$$

Substituting Eq. (11.1.17) into Eq. (11.1.14), the torque T is represented by the following equation:

$$T = 4\pi H \omega \frac{R_a^2 R_b^2}{R_b^2 - R_a^2} \mu_a \tag{11.1.18}$$

Therefore, the apparent viscosity μ_a can be obtained by measuring the torque T.

In the design of the coaxial cylinder viscometer, it is essential that centering be precise. Temperature control should also be provided. The surface of each cylinder should be roughened so as to ensure contact between the slurry and the surface. For particles larger than micrometer size, the imperfections in the surfaces of well-machined cylinders are not large enough to entrap the suspended particles and there is slippage at the wall. When dealing with large particles, a wider gap (10 to 100 times the diameter of the large particles) between the cylinders and larger protrusions must be used to catch and hold some of the suspended particles. As the suspension within the spaces between the protrusions will not flow, the protrusions should be included in the radius of the cylinder. End-effect or other causes for the measurement error should also be considered [7].

As a type of rotational viscometer, mixers associated with torque meters can also be applied in measurement of the viscous property of slurries. The measurements may not be rigorous,

but they are still advantageous enough for practical purposes.
Other viscometers [7], such as the vibrating-reed, float, and
plunger viscometers, may not be adequate for slurries.

11.2 MIXTURE QUALITY OF POWDERS

The measurement of powder composition is usually carried out
by use of an off-line instruments such as an x-ray fluorescence
analyzer after sampling the process powder. Qualities of powder
mixtures can also be evaluated based on a colorimetric technique.
The method may give satisfactory information even when the
electron microscopy fails to reveal significant variations in the
mixture composition. It has been reported that a spectrophoto-
meter linked to a computer gave a satisfactory quantitative
assessment of talc-pigment mixtures [12]. As mixing between
the fine pigment particles and the coarser talc particles proceeds,
the color changes progressively.

A simpler on-line optical method and its basis is discussed
below. Rays of light incident on a powder bed are reflected in
various directions (diffusive reflection). The reflectivity
R (unitless, $= I/I_0$) for a powder bed of infinite thickness may
satisfy the Kubelka-Munk equation [13]:

$$\frac{2R}{(1 - R)^2} = \frac{s}{\alpha} \qquad (11.2.1)$$

where s is the scattering coefficient of particles and α is the
absorption coefficient. For a homogeneous mixture of two
different types of powder, 1 and 2, Eq. (11.2.1) is modified
as follows:

$$\frac{2R}{(1 - R)^2} = \frac{s_1(1 - x) + s_2 x}{\alpha_1(1 - x) + \alpha_2 x} \qquad (11.2.2)$$

where x is the volume fraction of powder 2. If it is assumed
that powder 1 is white and powder 2 is black, the absorption
coefficient $\alpha_1 = 0$ and the scattering coefficient $s_2 = 0$. Then
Eq. (11.2.2) is simplified to

$$\frac{2R}{(1 - R)^2} = \frac{s_1(1 - x)}{\alpha_2 x} \qquad (11.2.3)$$

The reflectivity R = 1 for x = 0 (white powder), and R = 0 for x = 1 (black powder). The reflectivity is greatly reduced by adding a small amount of black powder. Equation (11.2.3) gives

$$x = \frac{1}{1 + 2R\alpha_2 / (1 - R)^2 s_1} \qquad (11.2.4)$$

Therefore, the volume fraction x of the black powder can be estimated by measuring the reflectivity R. As mentioned above, the reflectivity of white powder is unity, and magnesium oxide (MgO) powder is utilized as a standard material of R = 1.

For actual applications, the sensors should be calibrated by use of sample powders of known mixing ratios, because the reflectivity is also affected by other factors, such as particle size, particle shape, and packing density. The reflected light is detected by use of a photodetector such as a photomultiplier or a GaAsP-photodiode to give an electric signal. If the color does not change during the mixing process, colored tracer particles will be added, as shown in Fig. 11.4. These colored particles are separated after the reflectivity is measured. For constant quality control, it is usual to use a reference powder bed of desired composition. Only the difference between the output signals from the reference powder and the process powder is fed back to the process controller. The system is less affected by the possible fluctuations of the light source. Also, the system does not require the calibration.

An optical fiber probe can also be utilized as a quality sensor based on the reflectivity. The intensity of the light reflected from a powder mixture has been continuously measured by a photometer with five sets of optical fiber probes located separately in a batch mixing vessel [14]. Figure 11.5a shows the fiber probe. Two optical fibers of 1 mm diameter are set in a metal tube. Each tip of these optical fibers is expanded to 1.3 mm diameter and the end surface is polished. The metal tube is tilted at 45°, as shown in Fig. 11.5a and a glass plate is fitted.

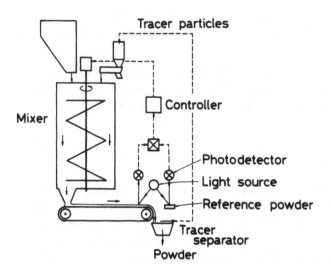

Figure 11.4   Quality control of a powder mixture based on an
optical method.

a)

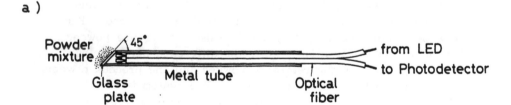

b)

Figure 11.5   Quality sensor composed of optical fibers and the
measuring system.

The outside diameter of the tube is about 6 mm. Light from a
LED (light-emitting diode) is transferred in one of the optical
fibers and emitted from the polished end into the powder mixture.
The light source (LED) is driven by an oscillator at a preassigned
frequency f_i, as illustrated in Fig. 11.5b. The light reflected
from the powder mixture is transferred through the other optical
fiber to the photodetector. The output signal of the photodetector
is treated electrically by an electrical amplifier, a suitable noise
filter, and a synchronous rectifier, giving an output signal of
the same frequency as that of the emitted light. The measuring
system has the feature that the detected reflectivity is not
affected by surrounding light sources, including another sensing
probe which may emit a light under a preassigned different
frequency f_j.

A microcomputer is applied to the data acquisition and
subsequent calculations. Then the degree of mixing M(t) defined
by the following equation is obtained:

$$M(t) = \sqrt{\frac{1}{n-1} \sum_{i=1}^{n} \left[\frac{x_i(t)}{x_\infty} - 1\right]^2} \qquad (11.2.5)$$

where n is the number of probes (= 5 in this case), $x_i(t)$ the
volume fraction or the concentration of specified particles detected
by the ith probe at time t, and x_∞ the expected value in the
final homogeneous mixture. The time-dependent degree of mixing
M(t) may offer valuable information as to control of batch mixing
process.

11.3 MOTION SENSORS

Various powder feeders are utilized in powder handling processes
as their final control elements. Powder feed rates will be controlled by these feeders according to the manipulating signals
given by process conditions in order to attain the satisfactory
control of the process variables which are measured by the
various instruments discussed in preceding sections. The feeders,
however, sometimes encounter difficulties arising from particle

adhesion, caking, or clogging. A shaft connecting the motor
drive and the feeder may be broken by the unexpected stress.
In such a case, particles may not be fed to the process even if
the motor drive is in normal working condition. Flow detectors
(described in Section 4.8) offer some help in avoiding such an
abnormal situation in the process. Sensors for detecting
rotational speed may also be applied. There are various speed
sensors based on the optical reflection. Optical methods have,
however, difficulty in keeping the light path clear against the
attack of fine dust. Some other speed sensors are based on the
phenomena described in Chapter 4: piezoelectric phenomena
(Section 4.5.2), electromagnetic induction (Section 4.5.4), or
spatial filtering (Section 4.5.5) [15]. The piezoelectric motion
sensor detects vibrations caused by the rotation of the motor
or feeder system through deformation of the piezoelectric crystal.
Although the output signal of the piezoelectric sensor contains
some noises, it is electrically treated and the periodic component
arising from the feeder motion is obtained.

The electromagnetic induction is applied in several ways,
such as in the dynamo or the eddy-current type. Figure 11.6
shows one of the methods for utilizing electromagnetic induction
to a motion sensor (Milltronics, Inc.). It consists of a permanent-
magnet rod and a coil wound around the rod. The permanent-
magnet rod forms a magnetic field around the rod as shown in
Fig. 11.6a. As for the differential transformer described in
Section 9.3, the magnetic field is formed not by a permanent
magnet but by a coil supplying an electric current. In the case
of the motion sensor, some of the magnetic flux lines pass through
the coil. If an object composed of ferromagnetic material passes
through the magnetic field, the lines of magnetic flux are distorted
as shown in Fig. 11.6b. The magnetic flux through the coil
will be changed by the distortion. As shown by Lenz's law, an
electric current will be produced in the coil to compensate for
the change in magnetic flux. The output voltage V caused by
the compensation is given by the following equation:

$$V = -\frac{d\phi_m}{dt} \qquad\qquad (11.3.1)$$

where ϕ_m is the magnetic flux [Wb (= V·s)].

(a)

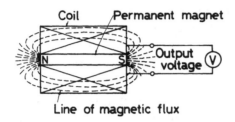

Line of magnetic flux

(b)

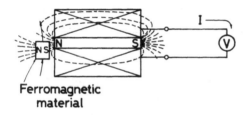

Ferromagnetic
material

Figure 11.6   Motion sensor based on electromagnetic induction.

As the magnetic field is formed by the permanent magnet,
the sensor does not require an electric source. The sensor can
be covered by plastics or ceramic, because the magnetic flux
lines are not affected by these materials. Stainless steel can also
be applied. If the trough of a screw feeder is made from stainless steel, the motion of the screw can be detected from the
outside even if the trough is full of particles. The screw blade
passes periodically through the magnetic field of the sensor, and
the rotational speed will be obtained through an adequate electric
treatment of the output signal. If a feeder has no such blade,
a piece of ferromagnetic material should be attached at a suitable
position of the feeder.

REFERENCES

1. R. B. Bird, W. E. Stewart, and E. N. Lightfoot, <u>Transport Phenomena</u>, John Wiley & Sons, Inc., New York, p. 10 (1960).

2. A. Einstein, <u>Ann. Phys. (Leipzig)</u> <u>19</u>, 289 (1905); <u>34</u>, 591 (1911).

3. K. Iinoya (ed.), <u>Funtai Kogaku Binran (Powder Technology Handbook)</u>, Nikkan Kogyo, Tokyo, p. 205 (1986).

4. M. Moony, <u>J. Colloid Sci.</u> <u>6</u>, 162 (1951).

5. Y. Mori and N. Ototake, <u>Kagaku Kogaku</u> <u>20</u>, 488 (1956).

6. E. C. Gay, P. A. Nelson, and W. P. Armstrong, <u>AIChE J.</u> <u>15</u>, 815 (1969).

7. J. R. Van Wazer, J. W. Lyons, K. Y. Kim, and R. E. Colwell, <u>Viscosity and Flow Measurement</u>, John Wiley & Sons, Inc., New York (1963).

8. Reference 1, p. 67.

9. T. Allen, <u>Particle Size Measurement</u>, 3rd ed., Chapman & Hall Ltd., London, p. 220 (1981).

10. B. H. Kaye, <u>Direct Characterization of Fineparticles</u>, John Wiley & Sons, Inc., New York, p. 102 (1981).

11. Reference 1, p. 89.

12. P. Bannister and N. Harnby, <u>Powder Technol.</u> <u>36</u>, 275 (1983).

13. K. Kubo, G. Jimbo, A. Suito, H. Takahashi, and S. Hayakawa (eds.), <u>Funtai-Riron to Ouyo (Powder-theory and Applications)</u>, Maruzen Co. Ltd., Tokyo, p. 388 (1979).

14. M. Satoh, Y. Deguchi, S. Komura, and K. Miyanami, <u>J. Soc. Powder Technol. Jpn.</u> <u>22</u>, 79 (1985).

15. H. Yamasaki (ed.), <u>Jido Seigyo Handbook—Kiki Ouyou Hen (Handbook for Automatic Control—Instruments & Applications)</u>, Ohmsha Ltd., Tokyo, p. 57 (1983).

Index

Milton Keynes UK
Ingram Content Group UK Ltd.
UKHW020023071024
449327UK00032B/2905